T0012510

THE
VISIBLE
UNSEEN

UNSEEN

Translated from the Spanish by
Kelsi Vanada

Artwork by Fabiola Menchelli

RESTLESS BOOKS
BROOKLYN, NEW YORK

ANDREA CHAPELA

THE
VISIBLE
UNSEEN

Translated from the Spanish by
Kelsi Vanada

Artwork by Fabiola Menchelli

RESTLESS BOOKS
BROOKLYN, NEW YORK

First published as *Grados de miopía* by Tierra Adentro, Mexico City, 2019

First Restless Books hardcover edition October 2022

Hardcover ISBN: 9781632063526
Library of Congress Control Number: 2022937665

Cover design by Emily Comfort
Text design by Tetragon
Cover image by Fabiola Menchelli

This book is made possible by the New York State Council on the Arts
with the support of Governor Kathy Hochul and the New York State Legislature.

Printed in United States

1 3 5 7 9 10 8 6 4 2

Restless Books, Inc.
232 3rd Street, Suite A101
Brooklyn, NY 11215

www.restlessbooks.org
publisher@restlessbooks.org

For my parents, the Mathematician and the Physicist

CONTENTS

Human beings have a curious spirit and poor eyesight. If we had better eyesight, we could see whether or not stars are suns which light their worlds, and if we were less curious, we wouldn't care. The trouble is, we want to know more than we can see.

BERNARD LE BOVIER DE FONTENELLE

THE
VISIBLE
UNSEEN

THE ACT OF SEEING

I DON'T REMEMBER how the discussion began—but I know it took place in January, when Chicago is covered in snow and it's impossible to be outside. From Iowa, I cross five hours of barren countryside in harsh weather to visit my friend A, who is getting her PhD there. We plan to eat dinner at the home of some mutual friends: two writers and a photographer. I think the discussion starts during dessert, when there's still a bottle of wine left. Someone (maybe me) makes a comment about some scientific topic, or about the vegan meal, or about A's thesis. Someone else (one of the artists) decides to play devil's advocate, wondering how science can be perceived as reliable (that is, "true") if theories change over time and often contradict each other. A and I try to explain why *precisely that* is the marvel of science, what separates it from dogma. The details of the discussion aren't important, and I've forgotten many of the arguments.

Suffice it to say that A and I are on one side, defending science, and the artists are on the other, casting doubt. More than anything, I remember my frustration. It was as if we spoke different languages and were incapable of communicating with each other. I tell all this to arrive at one particular moment: I'm searching for an irrefutable example, so I grab a knife from the table and drop it on the floor to illustrate gravity and Newton's laws. But I get all muddled up and fail to convince anyone. A interrupts me and explains what I just said over again. I'm uncomfortable. I know my lack of scientific rigor annoys her. For the first time, I feel that in choosing writing, I'm distancing myself from the world of science that has surrounded me since childhood.

Miroslav Holub, Czech poet and immunologist, wrote: "The sciences and poetry do not share words, they polarize them." As soon as I read that phrase, I wanted to negate it. Contradict it. It comes back to me when I set out to write a series of experiments—let's call them "essays"— to look at science through the lens of poetry and observe the result of that polarization: liminal language.

I met A in my first year of chemistry at the National Autonomous University of Mexico. Our friendship was always grounded in our similarities: we both studied pure chemistry and were fascinated by quantum chemistry. We both had scientific parents. We both read science fiction and fantasy. We had both gone to a public university after studying at private secondary schools. She played piano and composed music; I wrote novels. We were connected by our penchant for competing interests—on the one hand, science, and on the other, art—as well as our doubts about what to choose as a future profession. After four years, A started a PhD program in chemistry in Illinois, and I moved to Iowa to study creative writing.

"What is the name for the silence between the lightning bolt and the thunder?" wonders the Asturian poet Xuan Bello. I go to a poetry reading he's part of, and weeks later, I'm still turning this idea around in my mind. How can that moment be described? A lightning bolt is an electrical discharge that ionizes the air molecules. It cuts through the atmosphere in a matter of milliseconds, heating the gas and expanding it. The hot air increases in volume until it crashes into the surrounding cooler air currents. The air contracts. A rapid movement, violent. The silence between the lightning bolt and the thunder is the crack of a shock wave. No one word names the full process, but we are all familiar with the feeling of expectation. We hold our breath. The air expands, contracts. Rumbles.

Before I sit down to write, I seek out models to guide me, a process that's a holdover from my scientific education. I trust the clarity definitions can provide at the outset, and I feel it's easier to understand something if it's named. When I come across the category of "lyric essay," I cling to it. The genre gives me the form even before the foundation is settled and helps me organize my thoughts. In looking at science from a poetic point of view, I'm trying to encounter it afresh by making it unfamiliar.

Sometimes the process—not of writing a book, but of living it—reveals what we're looking for. Thinking of these three essays as one unit sends me back to my feeling of separation from science, to my friendship with A, to one poet's question and the words of another. I'm trying to braid all these threads together not to track the book's progress, but my own process as I create it. Lyric essay is an experimental genre because it's based on experience. And writing from experience means allowing the world to break into the text. This prologue is another experiment. Generated at the end, to be read at the beginning, it becomes a statement of intent.

I think back to A's interruption. It was like a lightning bolt, a kind of catalyst that made me realize how far I'd strayed from the concepts and language of science. From the ionization and expansion that followed, I gained clarity. I want to get close to science again, look at it with new eyes. That's why I must begin with this confession: to me, science is beautiful, and when I write, I want to explore that beauty.

Perhaps the pages that follow are simply an attempt to cling to all the parts of myself and never let them go.

THE ACT OF
SEEING THROUGH

Object of Study: Glass

1　I grew up in a house made of wood and glass. "Nothing to see outside. The house should look inward," said the architect, and in the center he built a garden.

2　When I want to talk about my past as if it were a story, I begin with: "In that house there lived a Physicist, a Mathematician, a Biologist, and a Chemist." Then it's easier to explain that what happened is the same thing that happens to all families. The Chemist left, abandoned chemistry. But can what's absorbed through osmosis ever be fully left behind? Answer: chemistry is the study of changes in matter.

3　The roof of the house is also made of glass. In the morning, the sun comes through its twenty-six panels, nourishing the two ficus trees in the indoor garden. At the end of August, it shines in my eyes as I sit there wasting time. I tell my parents that now that I'm in Mexico I'm going to start a book. But instead of writing, I'm sprawled in an armchair looking up at the roof. I look through it at the sky. I spend so much time looking at it that the object of my study stops being what I see through and becomes what I see. The glass.

4   A definition by Horst Scholze: "Glasses are undercooled solidified melts."

5   I'd be lying if I said that my mother, the Mathematician, and my father, the Physicist, tried to stop me from quitting chemistry. They knew that after four years of the study of matter, and twenty-five years of scientific cohabitation, I'd digested scientific thought and language. I was the only one who ignored this fact when I left Mexico for the United States to write. But there's no escape if what you find when you look inside is always science. Little by little its language seeped into my poems, and I started writing about bonds, synthesis, reactions, and decay. Hybrid words trapped between two worlds.

6   Linus Pauling, author of the first *General Chemistry* (1947), defined chemistry as the science of substances—their structure, their properties, and the reactions that change them into other substances over time.

7   Why build a roof out of glass? A skylight, maybe, but a whole roof? Glass is fragile, breakable, amorphous, cold, translucent, disordered, unreliable, and—as I discovered with my first boyfriend— doesn't allow for privacy.

8   Two months in Mexico. A static equilibrium between the United States and Spain. The days dissolve one into the next; they turn into one long day, longer than long, with no school and no work, reminding me of summer as a little girl or weekends as a teenager. I sleep in my childhood bed, surrounded by the dolls I collected and the books that returned home with me. My suitcase is in a perpetually half-packed state, and disorder radiates from it. Piles of books that will stay, notes that will come with me, sweaters, shoes, notebooks, tax forms. My surroundings reflect my circumstances: I'm in between two phases.

9   A few years ago, the roof leaked. All fall, the gutters had filled with leaves, and during the first storm of the season the rainwater trickled in between the glass roof and the stone wall. It turned the wall into a waterfall, the floor into a puddle, the interior into an exterior.

10 "A glass is an example, probably the simplest example, of the truly complex."

11 What is glass? (Consult entries 4, 26, and 57.) Even the most basic sources disagree. The Royal Spanish Academy (RAE): glass is a hard, fragile, and transparent or translucent solid without a crystalline structure. A delicate and breakable thing or a person of delicate temperament, easily irritated or angered. Colloquial Spanish expression meaning to take the blame: "to pay for broken glass." When I google "glass is a liquid": glass is a supercooled liquid, a viscous material that flows very slowly, so slowly that it would take hundreds of years to flow at room temperature. Wikipedia: common glass. Composition: silica, lime, and soda melted together at 1800°C (3272°F) and cooled until they form a disordered structure.

A material that doesn't behave like either a solid or a liquid.

12 In everyday speech, the words "glass" and "crystal" are interchangeable. At most, crystal is more expensive and more delicate. For example, my grandmother's wineglasses are made of crystal, not glass. Chemically, however, they're completely different objects. Crystals are solids whose atoms and molecules have a regular order that repeats in space. Table salt is a crystal: sodium and potassium molecules arranged in even cubes. Glasses lack this structure. They're chaotic, and we don't even know if they're solids.

13 In his introduction to *General Chemistry*, Pauling writes: "The words that are used in describing nature, which is itself complex, may not be capable of precise definition. In giving a definition for such a word the effort is made to describe the accepted usage." But when I used to say to the Mathematician "coming, ten more minutes," I wasn't thinking about how ten minutes is 600 seconds and one second is the duration of 9×10^9 periods of the radiation corresponding to the

transition between two hyperfine levels of the ground state of the cesium 133 atom. A minute is relative in everyday life, but in science, words aspire to mathematical precision, to a perfection that evades them in their fluidity.

14 In Mexico City, the city of forgotten rivers, everything that flows is piped, buried underground until it stagnates. Everything except the rain. The fat, cold, noisy drops revive the seven lakes and forty-five rivers. A decade ago, there used to be a rainy season, but now it's unchecked, so it pounds on the roof all year long. One afternoon at the end of August, about a month after I get back, the area around my neighborhood floods. The water reaches car windows, the Biologist is stuck and can't get home. From my bed, I listen to the storm, remembering my childhood fantasies. I used to imagine that one day the raindrops and the hail would bore through the roof. The crashing rain and shattered glass would pour in onto the garden. A river would form and flood my room, set my bed adrift. It would never occur to me to swim, so I'd steer along the newly created rivers until I ended up far, far away from home, with no

way to get back. The return of the water to Mexico City. As I write this, I realize it's not the first time I've described this fantasy.

15 The difference between a solid and a liquid is the difference between the two ways of fitting a set of billiard balls into a box: carefully, one by one, one on top of the other, in neat lines, creating order, compactly packed, a solid. Or, letting them fall with only chaos as their guide. When it comes to a box of billiard balls, my sense of order has always been liquid.

16 When I was a teenager, the Mathematician used to ask me to clean my room every three days. The perfect answer, "I can't. All spontaneous processes tend toward an increase in entropy, toward chaos," always made her smile. But it never worked: she would tell me that I just had to use more energy. In my house, applying the Second Law of Thermodynamics to my chores wasn't enough to get out of doing them. Now that I'm back home, my mother's insistence returns too, but after a longer interval of days, as if my age or our time apart have changed her standards.

17 The American Society for Testing and Materials
    defined the simplest method for differentiating a
    solid from a liquid in 1996 (ASTM D4359). Place
    the sample in a closed container at $38 \pm 3°C$ ($100 \pm$
    $5°F$) until it reaches thermal equilibrium (between
    18 and 24 hours). Take it out of the oven imme-
    diately, remove the lid, and invert it. If it flows
    a total of 50 mm (2 in) or less in three minutes,
    it's considered a solid; if not, it's a liquid. All my
    attempts to follow this methodology end in disas-
    ter. A puddle of water, a handful of rocks, a dribble
    of honey. They all end up on the floor.

18 An experiment. Stand at a window. Touch its
    surface with a finger. Feel the resistance. Now a
    second finger. Rest your entire hand against it.
    Push. Feel its solidity. Understand that a macro-
    scopic view is useless in this case. Imagine instead
    that it gives way beneath your fingers, liquefies.
    Would it be cold, wet, viscous? I forget about the
    experiment and rest my face against the glass.

19 Main characteristics of a solid: resists changes
    in form or volume, has a defined shape, par-
    ticles are closely packed and ordered. Main

characteristics of a liquid: has a defined volume regardless of pressure, but takes the form of its container. A cubic milliliter of water is the same in a cup, a bowl, a vase, the palm of my hand, a bathtub. And all those milliliters share the most important characteristic of a liquid: the ability to flow.

20 And glass? Does it flow?

21 The language of science today is English, but writing in Spanish makes things simpler for me given this particular subject. In English, glass is a broad category, in which everything made of this material is a *glass*. Maybe this expresses its deep instability more accurately than Spanish's many words. Not only is the threat of breaking glass ever-present, but a glass can be the thing for holding water *or* wine. Then there are eyeglasses for seeing, hourglasses for measuring time, sunglasses, magnifying glasses to enlarge small things and spyglasses to observe the faraway, storm glasses for measuring the weather, and even the glass formed by lightning at the beach. Glass is one thing, glass is all of them. In English, even

growing up under a ceiling made of glass would be fraught with multiple connotations.

22 "The deepest and most interesting unsolved problem in solid state theory is probably the theory of the nature of glass and the glass transition."

23 "Flow" is when atoms can be displaced easily; they're not tied to each other, they're not static. Fluids flow (it's a characteristic, not a redundancy—scientific language isn't afraid of repetition) because under any force, they transform and put up no resistance. The Mathematician used to tell me: "You're like a fluid: you adjust to your containers, you transform, you choose to go around the obstacles in your way." How easily the scientific becomes metaphorical.

24 Among the types of fluids are the Newtonians and the non-Newtonians. Before I knew who Newton was, I already understood these categories. When I was growing up, even the most mundane things could become the source of scientific explanations. At dinner, the Physicist used to take a jar of crema and shake it as he explained for the nth time

(science slips in with very little prompting) that crema is a non-Newtonian fluid: "If you exercise a force on it, it becomes more liquid, it's easier to pour."

25 I spend hours watching videos of glassblowing. Years ago, in a lab in Wisconsin, a PhD student showed me how to heat glass test tubes till they glowed bright red, deforming them to connect beakers to a vacuum pump, creating a many-legged lab insect. These asymmetrical creatures held a translucent liquid inside of them that slowly changed color, a magnetic stir bar spinning it for hours, a chemical reaction in process. Centrifuge for a day, let rest for a few hours, then break the insect open and extract a product that will later be used to kick off another reaction.

26 Supercooled liquids are partway between a solid and a liquid. Near the melting point, the molecules are moving, but run the risk of spontaneous crystallization. A glass is cooled beyond cold, beyond its freezing point, beyond solid, until the molecules have lost all possibility of movement: they're stuck between order and disorder in a metastable

state (consult entry 56). Christian Bök said it best in his poem "Glass":

> *Glass represents*
> *a poetic element*
>
> *exiled*
> *to a borderline*
>
> *between*
> *states of matter:*
>
> *breakable water*
>
> *not yet frozen,*
>
> *yet unpourable.*

27 September finds me working at the dining room table—not just because it's made of glass, but because I can spread out all the books, journals, and notes I need to write. During a family dinner, sitting at this very table, I can see my sister's long legs, my father's brown moccasins, my mother's compression socks. We used to have a dog that would jump up thinking the food he saw was for him, crashing into the table. They ask what I'm working on and I tell them about glass' aggregation dilemma. I explain that no one understands

what it is. My sister says she learned in school that cathedral windows are thicker at the bottom, due to gravity, because they're liquids. I look it up later. The windows of European cathedrals are made of blown glass framed in lead. At room temperature, the viscosity of glass is 1020 P and that of lead is 1011 P (how credible is this fact? does it matter?), which is to say that if a European cathedral window began to flow, drops of metal would fall first, not glass.

28   Pitch is a supercooled liquid, though other studies describe it as a brittle solid—in the end, it's more or less the same thing. In 1930, a sample of pitch was placed in a glass funnel at the University of Queensland. Since then, nine drops have fallen, the most recent one on April 17, 2014. It's estimated that the viscosity of pitch is 2.3×10^{11} times greater than that of water. It flows. Unlike the glass funnel that holds it, it can drip.

29   Every Christmas, my mother makes dulce de nuez using my great-grandmother's recipe. It's a complicated dessert to make, because it depends entirely on how the syrup is prepared. It looks easy, but

sometimes it takes a few tries. You heat water and one centavo's worth of sugar on low, and stir continuously. The sugar dissolves, the liquid thickens. My mother stirs; the sound of the metal spoon on the bottom of the pan fills the kitchen. Every so often she scoops some of the liquid out and lets it drizzle back into the pan, her eyes fixed on the stream of syrup. She's watching for the "thread stage" (when the thread divides into little drops, according to my great-grandmother's notes). She raises the spoon, inspects the syrup, stirs. She tells me it's getting close, watch carefully: "Look, look, *now*." The thread divides, the liquid thickens, drops form. My mother turns off the stove, removes the pan from the heat, still stirring. She says this time it will turn out well.

30 It's two in the morning. The site is called The Tenth Watch—the tenth watch for the tenth drop. As the video loads, a message appears: "Hi! Only 14 or so years to go." Then an image: the pitch sample, its glass funnel, the black drop suspended halfway. Everything under a glass bell jar. Each time I log on, I'm one of seventeen viewers. Who else in the world is watching? Is it a fake number, automatically generated to make me think I'm not alone? Will I be the

only human being alongside the computers supervising the drop when it falls? For a few moments, my imagination gets the best of me, and I think it moves, the drop widens, maybe it's almost there, it's going to fall, so close . . . but nothing happens.

31 Where does the myth about stained glass in Gothic cathedrals come from? Someone observed that a few pieces of glass were thinner near the top, as if gravity were slowly causing the material to flow, thickening the base. It's false evidence, still taught in schools. The difference in thickness is a characteristic of blown glass, a flaw in the process. A theoretical experiment was proposed to disprove the concept: suspend a plate of glass one meter in length and one centimeter thick at room temperature. How much time would it take for the glass to flow so that the base widened by 10 Å? The answer: the age of the universe.

32 If glass is not a solid and its fluid nature is a myth, then what is it? Confronting this question takes me back to my first poems, my attempts to blend chemistry and literature, how I wanted them to make both scientific and poetic sense. Why was I

so wrapped up in this goal at the time, and why am I still so wrapped up in it now? Does the amount of time a glass takes to flow matter more than the image of the windows melting, the drops falling, all the glass in the world transformed into water? Is the mystery of the nature of glass strong enough to hold all my worries and all the words I'm putting together? Is my urge to describe something unknown and outside the reach of language, enough? Glass is a destabilizer.

33 Imagine you're a melted material, boiled at over 1800°C (3272°F) until white-hot. You begin to cool slowly, so slowly that your atoms huddle closer together. You get heavy. On the phase diagram, you're moving toward a solid state, where you'll crystallize. Or not. Maybe you have a glass-like nature, and you can feel your molecules keep moving beyond the melting point, as you drop along the slope. Imagine the disorder inside you, growing along with your viscosity, growing as the temperature lowers to 20°C. You are almost a solid, but only barely.

34 The type depends on its composition. Glass composed with boron is used in labs because it resists

heat and doesn't crystallize; it doesn't break easily. The lenses of my glasses contain lead, and forest glass, produced in Cologne until the fifteenth century, is greenish due to its iron content. But 90% of glass is just silicon, oxygen, calcium, and sodium. The composition affects the glass transition (the moment on the phase diagram at which the slope changes) and instead of crystallizing, disorder begins and glass is formed. Systems at these points are very fragile; any change in temperature or pressure can alter the result. You get a liquid, instead of a solid. An experiment from university: cooling milk slowly, little by little, using cold water, ice, and salt. At -5°C, the milk is still liquid, but one tap of a fingertip and it crystallizes. In one second, it becomes a solid, unrecognizable.

35 I was in my third semester at university when I studied phase diagrams for the first time in my kinetics and equilibrium class. The basic example is the phase diagram for water. I can still draw it from memory. Two axes, temperature (x) and pressure (y); a line curving out from the origin to branch into two. The space on the far left above the first line indicates a solid, the space between

the two branches is liquid, and below the right branch is gas. The lines represent all the points of sublimation, melting, and evaporation: where water is simultaneously liquid and solid, liquid and gaseous, or solid and gaseous. But there is just one point along the axes of temperature and pressure, right where the line bifurcates, where water is a solid, a liquid, and a gas all at the same time. This point is found at 273.16 K and 4.65 mmHg. I look at my drawing, circling the triple point over and over again, wondering what the external conditions of the point would have to be for all my mental states to reach equilibrium.

36 Phase diagrams are graphical representations of the boundaries between the different states of matter of a system. As a liquid freezes toward becoming a typical solid, it goes through a phase change: the molecules line up little by little next to each other, one on top of another, in a simple pattern. They form a crystal. When a glass cools, the liquid becomes more and more viscous until it solidifies without ever attaining a rigid order. The molecules move slower and slower until they're trapped in a strange state between liquid and

solid. Is it possible that one single theory could explain all the kinds of glass? In the end, glasses are defined not by one common characteristic, but by the lack of one: the absence of order.

37 My favorite experiment: I'm in my kinetics and equilibrium lab, looking at a creature made of a glass flask and a rubber hose. A transparent liquid rests in its belly: 100 mL of cyclohexane, a hexagonal molecule that looks like a honeycomb. The boiling flask, suspended by a clamp, is connected to a vacuum pump and a thermometer. As the pressure is lowered, the temperature goes down too. We're looking for a point at 45 mmHg and 6°C. Lower, lower, lower—the cyclohexane bubbles, but doesn't boil like water does. The bubbles slow little by little, the liquid gets more viscous, the surface solidifies, and the gas turns into ice with each bubble that reaches the surface. The system comes to life. Inside the flask, a solid sinks, a liquid bubbles, and a layer of gas solidifies and forms droplets. At the triple point, cyclohexane turns opaque and is—at the same time—all three states of matter.

38 After my Spanish visa appointment, all I can do is wait. To distract myself, I talk with old friends and tell them all the interesting facts I've learned about glass. I become monothematic, full of "Did-you-knows?" I ask a friend: "Did you know that cotton candy is a glass?" He responds by telling me that glass breaks because the force crystallizes its edges. So I buy a drinking glass because my mother would get mad if I started breaking her glassware (our current cohabitation, since my return, feels too fragile to experiment with). When I tell her what I plan to do, she says that for the sake of art, it wouldn't have bothered her if I'd broken a glass. She advises me to shatter it inside a plastic bag or two for safety's sake. I pause multiple times before dropping it. It's strange breaking something on purpose, with a purpose. The first time the impact sounds dull and flat: the glass just bounces. The second time, I hurl it: again, a muffled sound. The third, the bag above my head, much more force, it has to work this time: and there, the sharp ring, crystalline, of something breaking. When I pick up the bag, I notice it's warm.

39 In the book *Líquidos exóticos*, the authors speak
of the eventual crystallization of glass: "It's estimated that some glasses, like the leaded glass in
Gothic cathedrals, will take thousands of years to
crystallize. When this happens, the glass will be
transformed into a crystalline solid and will break,
like car windshields, for no apparent reason." My
nighttime terrors could become reality (consult
entry 14). The roof of my childhood home could
break instantaneously, for no reason, just because
each panel will crystallize over thousands of years,
and the weight of order will shatter them.

40 No state is more metastable than waiting.

41 I open the bag in the outdoor garden, and gingerly
take out the fragments one by one. My fingers
are covered in a glassy dust. I observe them carefully, but gain no clarity. What was I hoping for?
How did I imagine crystallization would look?
The edges aren't translucent, I can't see through
them, and it's as though someone had polished
them. I can see waves, cracks. Crystals are solids,
but these edges are the first thing that makes me
think maybe glass *is* a liquid. Later, I look online

to see if what my friend told me is true. I learn that glass, in its disorder, has micro-cracks, and these invisible cracks make it fragile despite its durability. How glass breaks depends on chance, on the impact, on the force, on all the micro-cracks of the cooling process. Not a single web page talks about the crystallization of the edges, but does it matter? I observe one of the glass fragments, now formless, useless, more liquid than when it was whole. I know science needs precise, testable explanations, but do *I* need them? Or is this writing exercise—based on the reappropriation and transformation of my own language—enough? Do I need to find some kind of truth? Or are the edges of this debris—wavy, fluid, splitting the light to form a tiny rainbow in each crack—enough?

42 On September 18, my visa gets denied. The next day, everything breaks.

43 What breaks: teeth, backs, the peace, the silence, childhood friendships, plant stems, hearts, voices, ancient vessels, my dreams of working in research, ionic bonds, necks, hips, clavicles, the whole body, dawn, the news, stories, falls, my grandmother's

wineglasses, summer romances, run-on sentences, habits, records, health, faith, hair, social fabrics, pinky toes.

44 I'm in bed reading when the floor loses its solidity and ripples. In one second, all action is imperative reaction: jump out of bed, cross the indoor garden, get out of the house, look for your sister, call out for her, keep breathing, wait. Dogs bark, birds chirp, and humans keep quiet. The glass roof creaks, shudders, vibrates with each undulation. I'm certain it's going to break. But, when the earthquake passes, we only find that the quartz geode has tumbled off the shelf and shattered on the floor. The glass roof, on the other hand, didn't break.

45 How can I write about the cracks, between the cracks, when even language is in ruins? Not everything that breaks crystallizes.

46 Normal life shatters too. I stop writing, lose the thread of my thoughts and all interest in glass, language, or finding my place in my childhood home. I stare at the TV with my father, the rest of the rooms in

darkness and silence. An infuriating succession: buildings collapsing in clouds of dust and uproar / cries and banging from beneath the debris / pairs of dogs / young people in human chains passing empty buckets that return full of stones / pets abandoned and found / tables, backpacks, homework, dolls, frying pans, the signs of hundreds of lives / dishes full of food / the metro running for free, but its cars empty / people helping and waiting, people searching for a place to help or wait / messages asking for flashlights, helmets, medicine, water, and comfort. The fragmented images are superimposed during our vigil, as if the power of the gaze—televised, collective—could restore order.

47

48 In October, after a lot of paperwork and back-and-forth, I get my visa. After I buy my plane ticket, the days speed up. I start packing. I put photos, letters, and postcards—reminders of my family members and friends—into a metal box that will travel with me. I pause on a postcard I received last summer. It has on one side, "I hope you'll visit Madrid," and on the other, a picture of the

Palacio de Cristal. An epiphany: that's where this book is headed, toward Madrid and the Palacio de Cristal. That's the ending: the writer changes countries and finds her musings-turned-landmark. What could be more satisfying than architectural and narrative growth, going from a glass roof to a whole palace? I'm relieved; now I can think of my departure as a destination.

49 Knowing I'll be traveling soon makes me take up writing again. I read *Glass (Object Lessons)* by John Garrison. I admire his attempt to track glass through the depictions of the past and the imaginations of the future. I write down a quote: "Even when it's transparent and trying its best to be invisible, it's still affecting how we experience what is beyond it." He's talking about glass, but this idea could apply to all of language—scientific language, to be precise. How can I write about science from outside it? How can I stop seeing *through* language, using it as a tool, pretending exactitude is possible in words? What happens to scientific words when they're observed? If we extend the metaphor, we'd say they become unstable and change aggregation states.

50 My first few days in Madrid go by in a blur of the worst jet lag of my life. It takes me a week to accept that I've finally arrived, to unpack my suitcase and begin to put this new room in order. Though it's smaller than my room at my parents' house, I have my own bathroom, a big desk, and a window that looks onto the garden, with the city in the distance. I take out my notes, sit in my new chair, inspect my surroundings, but I don't write. Through the window, novelty beckons, and I get distracted.

51 I plan my first visit to the Palacio de Cristal carefully. I go alone, walking through El Retiro on a day when my being there has finally begun to feel routine. I don't go in, because the line to see the exhibition snakes around the corner of the building. I sit down to write on the other side of the pond. The Colombian sculptor Doris Salcedo called her intervention in the space *Palimpsesto*, a memorial made of water—of invisible, fluid words. In this case it is a "funeral oration," a poetics of pain and a way to mourn all those who have lost their lives trying to cross the Mediterranean. Salcedo asserts in an interview: "The future is built on the ruins of the past, and art helps clarify this

call to attention." That afternoon, when I return for dinner, I find out that Doris Salcedo visited the Residencia de Estudiantes; she met with my fellow residents while I was still in Mexico.

52 Over the course of the next few weeks, I return again and again to the Palacio de Cristal in search of an ending.

53 Doris Salcedo, who calls herself a maker of objects, was the first woman invited to take part in the Unilever series, intervening in the Turbine Hall at the Tate Modern in London. The museum's director said of the work: "There is a crack, there is a line, and eventually there will be a scar. It will remain as a memory of the work and also as a memorial." When she cracked the floor of the room, Salcedo wanted to represent borders: migrants' experiences of crossing them, as well as their destinies upon reaching the other side.

54 For Salcedo, the crack is negative space. Everything fits inside it and is "too many things," as Borges writes in *Other Inquisitions*, of the chasm that appears in the Roman Forum in Hawthorne's

The Marble Faun. "It is the crevice mentioned by Latin historians and it is also the mouth of Hell *with half-shaped monsters and hideous faces*; it is the essential horror of human life; it is Time, which devours statues and armies, and Eternity, which embraces all time."

55 I leave the Palacio de Cristal without knowing where these reflections are taking me. I was hoping for an epiphany, but I only find metastability. I wonder where the equilibrium is when you no longer belong to your childhood bed, or to your girlhood, or to the bedrooms in the countries you choose as resting points.

56 Was using the word "metastability" over and over without pausing to define it an oversight? (Consult entries 26, 40, 55, and 59.) We can imagine the evolution of a system over time as a marble rolling along a track that goes up and down, through valleys and over mountains. The marble is in equilibrium when it is located at a point of kinetic stability (that is, in a valley) where it can stay put until a push of some kind (a force, a change in temperature or pressure) gets it moving again.

At the end of the track (the ground state) when the track is level with the earth—that's the point of total stability. All the rest of the valleys are metastable states, always at risk of change.

57 In 2016, during the conference of the Society of Glass Technology at the University of Sheffield, Edgar Dutra Zanotto gave a keynote speech titled "Glass Myths and Marvels." He proposed a new definition of glass: "Glass is a nonequilibrium, noncrystalline state of matter that appears solid on a short timescale but continuously relaxes towards the liquid state. . . . [Its] ultimate fate, in the limit of infinite time, is to crystallize." It's the most recent definition published to date. Scientists circle around the idea of glass, refining its definition word by word, as if they were progressing millimeter by millimeter.

58 To pursue science is to assume that each repeatable experiment and proven hypothesis brings us closer to some absolute truth. At its core is the conviction that one day we'll be able to understand everything around us. When I studied chemistry, I developed the bad habit of searching for precision in words,

but my mistake was in forgetting that language is an approximation. Like believing that when I can see my breath in winter, I'm observing an ideal gas. I thought words were solid, reliable, but the exercise of writing has taught me that they mold to whatever container I put them in. They *flow*.

59 Glass, because of its metastability, is an orphaned material. This is due to the limitations of our language, the strictness of its taxonomy. Definitions in scientific language can't be fluid, yet faced with the mystery of glass, we have to accept the fragility of words, their lack of precision. Accepting this opens the door to searching for a way of talking about the most elusive experiences—the sensations and feelings that can only be grasped through metaphor, though we often fail in our attempt to capture them in language. In failing to define glass, in having to make comparisons and create new categories, I discover that the orphanhood of glass is also, in its turn, the fundamental failure and the very orphanhood of writing.

60 I'm told that the day I arrived in Madrid was the first cold day since summer ended. In the morning,

as I cross El Retiro, the chilly air feels crystalline, and under my feet the leaves crunch. I no longer need a map to find the Palacio de Cristal. I amble through it, more occupied with searching for my own reflection in the windows than studying the names on the floor. Unlike my house, the Palacio was built for gazing out: the pond, the blue sky, the trees in the park. The glass magnifies the birdsong. In Mexico, I always looked outward. But now, in Madrid, I'm looking inward. I pause before one of the walls, consider my reflection, and think about how I always have to go against the flow. So I focus on the glass and on the idea that one day everything around me will crystallize, will shatter, and in doing so will find equilibrium. Will I find it too? It's hard to determine the final state of the system when you're halfway through the process. Between two states is a transition, which sometimes reaches equilibrium and sometimes stays metastable—it always depends on the circumstances. Only with the passage of time, by looking back, can you tell which was the path.

61 Some days, I'm just relieved to remember that Lavoisier's law holds true for all of nature.

THE ACT OF SELF-SEEING

Object of Study: Mirror

I COULD BEGIN BY saying that mirrors are useless if no one looks in them.

Or: the history of mirrors is the history of (self-) contemplation.

I COULD BEGIN WHEN I arrived in Madrid. Not the moment I landed at Barajas airport, and not the first few weeks when the day-to-day still felt like a vacation. Begin then, with my real arrival, once the initial excitement is behind me and I find myself in a new country, a new bedroom, a new routine. I may not want to, but I think I need to reinvent myself in this place, or at least get to know myself again. *Who am I here?* For weeks, I'm uncomfortable in my own skin. I search for my reflection in every surface, making sure my feelings haven't broken out like a rash. I remember one of Idea Vilariño's poems. She wants to buy a mirror, hang it in the bathroom, and look at herself: "How else can I find out who I am?"

An experiment:

Go into the bathroom. Close the door. Turn off the

light. Search for a reflection. Fail to find it. Open the door. Let in light from the hallway. The face appears in sections. It takes shape little by little. First, an outline. Then the dark frames of the glasses. Finally, the nose and eyes. Touch the surface of the mirror. Note the space between the hand and its double, the chill of the glass. Close the door. Watch the self disappear. A thin strip of light filters in underneath the door. Eyes wide open. In the darkness, look for signs of the reflection. Feel its presence.

I could begin like this: facing the mirror, searching for my own countenance, I let the light in.

I COULD BEGIN WITH the first mirrors in the world. A riverbank, a puddle after the rain, a bowl of water. No contortion allowed the full body to be viewed; to lean too far over the edge was to risk falling in. Later, with the advent of the first tools, people polished fragments of obsidian until the shadows gave forth their reflections. The Egyptians regarded themselves in bronze mirrors. The Chinese in mirrors of jade and metal. The Greeks in round metallic mirrors with handles and embellishments. In the span of six thousand years, humanity went from rock to polished metal, from obscure to opaque. Until, in the first century BC, the first silver-coated mirror was made in Anatolia—but its surface wasn't even, the reflection wasn't perfect.

I COULD BEGIN TODAY: December 23. I meet up with a former roommate from Iowa, a poet, at La Duquesita, a pastry shop near Alonso Martínez that's been open since 1914. The shop and its desserts do credit to its name, "the little duchess." Its walls are white, and its tiny black and white checkerboard floor tiles are laid out in diamond patterns. In two glass display cases, miniature brownies, fruit tarts, and cakes with sugar icing glisten under the lights. There's a long line, but we have our hearts set on coffee and croissants, so while my friend waits to order, I grab a table on the left side of the shop. I amuse myself by describing everything around me. At the back of the room, there's a large mirror in a thin frame made of light wood. Two more mirrors—one at my back, the other facing me—flank the room, making the space seem larger than it is. I twist around to take in my own face, and then the reflection of the candelabras and the

51

line of people moving forward little by little. Finally, I study the mirror itself. It's riddled with black spots, as if someone had attacked it with wild brushstrokes. I notice how these oxidized stains afford the mirror, the whole pastry shop, a veneer of age.

I take pictures of the spots; my reflection observes them from a corner of the frame. I think of the other mirrors I've seen around Madrid, covering the walls of Café Barbieri and other locales my friend deems "bars of yore," and how the dirty, stained mirrors in these places give off an air of neglect and antiquity. Does time slowly eat away at them? Is it our collective destiny to turn back into a dark surface that absorbs light instead of reflecting it?

I COULD BEGIN WITH the Lighthouse of Alexandria, one of the Seven Wonders of the Ancient World. For a thousand years it guided vessels through the Nile. At the top of the marble and glass tower was a metallic mirror reflecting the sun by day and a fire by night. The ray it cast could be seen from fifty kilometers away.

I could describe Archimedes' heat ray. He was born in the second century BC, exactly one year after construction on the lighthouse began. Later writers tell us that during the Roman invasion of Syracuse, Archimedes used a burning mirror to reflect sunlight, focusing it into a ray that set fire to the invader's ships.

Or I could describe Aristotle, who wrote that if a woman looks into a polished mirror while she's menstruating, a reddish smoke will cloud its surface. The next time I'm on my period, I eye myself in the mirror defiantly.

I could admit that I'm overwhelmed by all these interesting facts and the challenge of narrowing down my research topic. I rotate between the present and the past, the experiential and the theoretical, the poetic and the scientific. Something at the center of these revolutions escapes me, even as I try to focus on it. I remember that this is not a historical essay, nor a treatise on physics, nor an autobiography. It's the search for a concave surface where I can collect my thoughts into a ray that sets fire to whatever it touches.

I COULD BEGIN WITH my first mirror, a glass rectangle with an aluminum coating. The wavy shapes carved into its reddish wooden frame give it the effect of being covered in leaves. I regard myself in it. Standing, I can only see my reflection from the waist up. To see from the waist down, I have to balance precariously on my rolling desk chair. I bought the mirror on a trip I took to San Miguel de Allende when I was fourteen. All my savings in return for the privilege of looking at myself. I'd like to be able to recall my teenage desire in detail so that I can record it, but the object alone remains.

Six months after buying the mirror, I came home from school and went straight to my room. At first, I didn't know what was wrong. I went back out and reentered a few times—then realized my mirror wasn't there. It used to hang between the bed and the desk, lined up with the door. When I asked what had happened, my mother answered that the mirror had fallen that morning. The frame was intact, only the glass had broken. She said she'd take it to get it fixed.

The mirror reappeared the next week. I realized it was there for the same reason I had noticed it was gone. I had become accustomed to seeing myself as I came through the door, crossed to the bed, dropped my backpack. The day my mirror disappeared, I didn't miss the object: what was absent from the wall was my reflection.

I COULD BEGIN WITH a painting. A nude woman reclines on a bed. Her back is turned to us, and she's looking into a mirror held up by a cherub. We see her face in it and assume she's gazing at herself. The painting is Diego Velázquez's *Venus at her Mirror*. It plays with our perception, makes us forget that if we can see her face in the mirror, she can't be looking at herself, but rather at the painter: that is, at us.

I don't know how old I was when I fully grasped this phenomenon, but I can remember learning it. I remember putting it to the test in the car, waiting for physics to fail. I focused on the Physicist's face in the rearview mirror, sought out his eyes, and asked if he could see me too. He nodded with a smile and explained yet again that when light is reflected off a surface, the angle of incidence is always equal to the angle of reflection. I recall the feeling of imagining my reflected face as he must have seen it, wondering whether it was the same as the one I saw each morning.

When it comes to mirrors, it's impossible to look without being seen.

Many years later, I find myself thinking about the Venus effect again. I'm taking a taxi along Viaducto Tlalpan. I requested the cab by phone, so it's theoretically trustworthy, but that's not enough to chase away my paranoia. I carefully observe the driver's face in the rearview mirror, calculate the probability that he'll harm me, and rehearse various escape routes in my mind. As I run through all the advice I've ever read, I grip my cell phone, ready to call home. I'm scanning for the tiniest gesture to justify my nervousness. I remember too late that, if I can see him, the driver can see me. He could hold my gaze, realize what I'm thinking, or worse, monitor me while I'm distracted. I shift slowly until my reflection is visible. In addition to my unease, I also feel ashamed for forgetting something so basic.

I COULD BEGIN TODAY: Christmas Eve. The Poet invites me over for dinner with her family. Ever since my fellow residents left Madrid, my routine has been the same each day until now. I'm not in the holiday spirit, maybe because it's the first time I've spent Christmas without my parents, or maybe because they don't decorate the Residencia. When I get to her house, the Poet takes me to her childhood bedroom, and the first thing I notice is that the doors of her closet are one continuous mirror. She plugs in a string of Christmas tree lights and sits down on her bed; we stare at our reflections silently. A year ago, we were living together: a door led from her room to mine, and we shared clothes and mirrors. We were so comfortable around each other that we didn't censor our thoughts when we got ready to go out, but listed off our insecurities aloud: big thighs, low-cut shirts, the dark circles under our eyes, the length of a dress.

I want to talk about the first mirrors that find their way into our rooms when we're little girls. I want to talk about the moment when we go from just wanting to look at ourselves to judging ourselves. I want to talk about the people who go on mirror diets—they cover up or hide all the reflective surfaces in their homes, learn to put on makeup by feel, and stop observing themselves for days or months. They say they have more self-confidence right away. I want to talk about the fact that the majority of my female friends feel self-critical when they look in the mirror. I want to talk about how the Poet's younger brother doesn't have a wall of mirrors in his room. I want to talk about these things, but I don't know whether I have anything new to say.

I COULD BEGIN WITH the steps involved in making a mirror. First you need a sheet of perfectly polished glass; any imperfection will distort the image. The sheet of glass is covered with a layer of silver or aluminum, heated to its boiling point and allowed to vaporize onto the surface. The trick is to create a uniform covering. In 1835, Justus von Liebig invented the process of silvering, ceasing the production of mirrors coated in mercury and lead. Silver turned out to be a very good reflector, but the problem was that it had to be sealed by a layer of protective backing. In 1930, the procedure for using aluminum was finally discovered, and that's why most mirrors today are made from this much more stable metal. The decaying mirrors at La Duquesita and many other bars in Madrid must be made of silver. When the material insulating a silver mirror's coating is damaged or aged, humidity can pass through it. The water molecules then react with the metal and corrode it. I scribble a simple equation in my notes. It's satisfying to balance, and I'm proud I haven't forgotten how to do it. $Ag_2NO_3 + H_2O \rightarrow Ag_2O + 2HNO_3$. I circle the Ag_2O, silver oxide, and imagine my mother sitting at the kitchen table with the box of my grandmother's tarnished cutlery. With a cloth, she cleans each piece until it's shining again.

I go back to La Duquesita and find myself surrounded by its mirrors once more, in the same spot as my first visit. I observe their imperfections. Now I understand what they are, what they're doing there, and what they mean. I sense a newfound connection with Madrid's bars—it's as if, now that I've grasped this fact, I've suddenly earned the right to inhabit not just this place, but the whole city.

I COULD BEGIN TODAY: December 29. On weeknights, the streets between Lavapiés and La Latina, where El Rastro market is set up on Sundays, are practically empty. I run into a few kids playing ball in a plaza; people with lowered eyes hurrying home from work; and groups of men shouldering their blankets full of handbags and shoes to sell, talking in languages I don't understand. I walk in circles. I avoid taking out my phone, and instead explore the narrow streets as I try to reach the Tirso de Molina plaza, unsure which direction to take. Far from Mexico City, I relish the privilege of walking alone at night. I stop to take pictures, peer into dark display windows, enjoy myself in each plaza. Between Christmas and New Year's, Madrid lies under a strange calm.

As I turn around in yet another steep, narrow, curving street, I come across some broken crates and two shattered mirrors. I eye myself in the pieces still leaning against the wall. Their cracks fragment my image: my legs in one segment, my arm in another, half of my face over there. The Romans believed that mirrors revealed your soul, and that, if you broke one, you also splintered a piece of yourself. The seven years of bad luck ascribed to whomever broke a mirror were the number of years it took the Roman soul to heal. I wonder what superstition they would have about a nail that gives way and a mirror that breaks on its own. Maybe the confusion I felt when I got home from school that day to find my reflection missing was, in fact, the reaction of a person with a cracked soul. Did replacing the glass repair it? What gets lost in these instances?

There's something disquieting about a broken mirror abandoned in the street. A personal attack.

I COULD BEGIN WITH Jacques Lacan, who described how children react when they first encounter their own reflection. He gave the name "mirror stage" to the phase of an infant's psychological development when they begin to recognize their own likeness. Children react to mirrors with glee, but this first self-identification, Lacan tells us, is imaginary. Without the help of reflections, seeing our full image would be impossible, and this affects how our identity is formed.

Philippe Rochat proposed that self-awareness is developed between the ages of six months and five years, in five levels: that is a mirror; in it is a person; that person is me; that person will always be me; what I see is the same thing others see when they see me.

Some mornings, I suffer through this entire process

again. *I am that person*. What I see unsettles me. *That person will always be me*. I don't recognize her, she doesn't fit my vague idea of who I am in my mind. *What I see is the same thing others see when they see me*. It seems so unfair that other people know what I look like better than I do.

Lacan says the dissociation we feel when we look in the mirror comes from the way we perceive our own likeness in pieces. Hands on a keyboard, a torso at a desk, legs disappearing under the table. This close to myself, I can hardly grasp the whole I make up. Every idea I have of myself is imaginary, which makes me feel helpless. I'd like to think I know myself, but even in this most basic aspect of who I am, I come up against an obstacle. So I think through all the mirrors I've ever encountered, and their different versions of my reflection. I try to endow the experience with fresh significance. As I'm writing, I sometimes lose the thread of what I'm trying to articulate, question whether it's working, and feel the need to draw on an external gaze. With the same vulnerability of someone asking, "Do I have food in my teeth?" or "Does it look like I've been crying?" I ask, "Do you see this text as I see it?"

I COULD BEGIN WITH a chimpanzee in front of a mirror. He touches the glass, observes his nostrils, sticks out his tongue, shows off his genitals. He rubs the spot of white paint on his forehead until it's gone. He has passed the mirror test: he is conscious of himself.

Orangutans, dolphins, killer whales, elephants, capuchin monkeys, rhesus macaques, magpies, and even ants look in the mirror, understand that they are the image they see, and try to rub off the mark the researchers have put on their bodies. Before the age of a year and a half, human babies react to mirrors with fear or curiosity; they don't associate the person they see with themselves. Gorillas will sometimes look away, cover themselves in shame, or attack the mirror—and fail the test.

The ability to see oneself in the mirror is as close as we can get to measuring humanity. But what about dogs, whose sight is worse than their sense of smell? Does recognizing the scent of their own urine, identifying an "I was here," make them conscious of themselves? Recognizing our own likeness isn't proof of the awareness of self. Maybe it's all nonsense, and it doesn't matter. But then again, despite its flaws, it's the only experiment that tests whether other species experience consciousness.

I still can't forget the feeling of looking at myself in the mirror when I was twelve, recognizing this taller, curvier, changing body as my own. I did the same thing not long ago, the day before I had surgery to remove a small lump of fatty tissue from my left breast. Standing in front of the mirror, I thought, "Take a good look, remember yourself at twenty-six, before you got a new scar." Is this vanity, or is it self-recognition? Why do I feel the need to withdraw into myself when I'm sounding out a new environment? Does it comfort me to open my closet door, regard myself in the mirror hanging there as Vilariño did at the age of eleven, and tell myself *here I am*?

I COULD BEGIN TODAY: January 1. At night, the window in my room turns into a mirror. From my seat, I study my reflection, punctuated by the lights of the city. It's a partial reflection: I can see myself and beyond myself. It's also a double reflection: my image and its shadow, because of the double panes of glass protecting me from the winter weather. Experiments show that 4% of light always bounces off glass, while the remaining 96% goes through it. Photons from my desk lamp hit my face, some bounce off it toward the window, and 4% (always 4%) return to my eyes.

Feynman begins his book *QED: The Strange Theory of Light and Matter* with a warning: he will tell us *how* nature works, but we can't understand *why* it behaves in the way it does. We're in good company: no one knows.

We know *how* photons act, such that our measurements always give 4%, but we don't know *why*. Likewise, when we look in the mirror, we recognize ourselves, but we don't understand *how* or *why* we do. As I study myself in the window, I think how much more useful mirrors would be if only they allowed us to see beyond—but even so, like Idea Vilariño, I see my face and believe it's the only possible one. All my internal images are erased, and I'm left with just my face speckled by the lights of Madrid. *That is me.*

I COULD BEGIN WITH a study published in the *Journal of Personality and Social Psychology*. It concludes that people are more productive with a mirror in the room, because if they're conscious of themselves, of occupying a space, they're more likely to pause, consider what they're doing, and improve their performance. A mirror in a room has the power to capture the gaze. This happens to me when I catch sight of myself in car windows, or in the reflections of storefronts as I walk by. If I write in cafés with lots of mirrors, I don't stare at them constantly, but I'm conscious of my body's copy in the periphery of my vision: its movements, its outline.

I remember the afternoon I spent with an ex-boyfriend in a hotel room with a mirror on the ceiling. I felt uncomfortable every time I noticed my reflection. I remember my aversion, how quickly it distracted me from what I was doing to see a bit of my skin out of the corner of my eye. I turned my back to my double, ignored it, tried to forget it was there.

When I tell the Poet about this, she laughs and says she likes looking at herself in the mirror when she's having sex. I'm surprised by my own modesty, and I can't seem to explain that it wasn't my body that bothered me. The repulsion I felt came from having to confront my own image in a moment that could have been pure sensation, not contemplation; I would've liked to forget myself entirely. I'm always trying to ward off my own self-consciousness—it inhibits me, pulls me out of the present, distracts me, and forces me to think instead of feel.

I COULD BEGIN WITH the mirror I bought when I moved to Iowa, the first time I lived alone. That's when I learned how easy it is to alter a reflection. All it took was propping my full-length mirror at an angle, not just so that it wouldn't fall, but so my body appeared taller and thinner.

An easy way to modify mirrors is to curve them. Concave mirrors curve inward, collecting light and reflecting it so it converges at a point. They're used in telescopes for observing stars, and in bathrooms as magnifying mirrors for putting on makeup. Convex mirrors curve outward, toward the light, enlarging the field of vision and distorting object size, like the side-view mirrors on cars. But since they're mounted inside their covers, we never remember that "objects in the mirror are closer than they appear."

I COULD BEGIN TODAY: January 2. I turn off the lights in my room, light a candle, and sit on the floor, facing the mirror on the inside of my closet door. I start a timer and stare into my eyes. It's hard to sustain my gaze. Time and time again I surprise myself, suddenly realizing that I'm observing my nose or the reflection of the light on my glasses. I take a deep breath and restart the timer. I concentrate on counting my eyelashes to make myself look into my eyes. They're dark, small. Suddenly, I lose focus, and my face blurs. I see shadows out of the corner of my eye. I get vertigo. I break my gaze and fix it on my hands instead, but the churning in my stomach doesn't go away. I close the closet door, stop the timer, and turn on all the lights in the room. To calm myself down, I remember that the visual system is conditioned to recognize faces. That's why it's so easy for people to see the Virgin Mary or Jesus in the clouds, in trees, or in cracks in the wall. In dim light, edges are muddied, so our specular face gets blurry. Even though I know this, I don't try the experiment again.

Any distortion in a mirror alarms us. If I raise my left arm, my double raises its right arm. The brain interprets our reflection as vertically inverted because our body is symmetrical in that direction, but that's incorrect. The inversion is from back to front. The reflection is a negative, a projection through ourselves. Just as light shining through a sheet of paper allows us to see what's written on the other side, but in reverse, our reflection is an impression of light on the mirror. This can be corrected by placing two mirrors perpendicular to each other, allowing us to see ourselves as others see us, without inversion. In 2015, a Spanish company announced that it had begun marketing this type of "real image" mirror, enabling users to see themselves in three dimensions.

More than once, I consider buying one online so I can experience what it feels like to *really* see myself. But the memory of my experiment stops me, the vertigo I felt when I studied myself by the light of a candle, and I wonder if the sensation would be even more unsettling if I were seeing a reflection that showed my real image.

I COULD BEGIN WITH a photograph. I'm standing on the threshold of the Hall of Mirrors in Versailles. Behind me are a bunch of tourists. Although my family and I had gotten lost in the fog trying to find our way to the palace that morning, it was sunny when we took the photo. At the time, I didn't know that the hallway was not only a show of Louis XIV's opulence and power, but also proof that the French had managed to steal the island of Murano's secret method of making mirrors.

For a hundred years, the council of Venice had defended their monopoly; the best glass and the best mirrors came from the island of Murano. But Louis XIV wanted to surround himself with mirrors. His ancestors had not had the luxury of looking at their full bodies in a mirror, and the closest thing to a reflection was a portrait. On my first trip to Paris, I insisted on posing for a portrait from a street artist at the base of Sacré-Cœur, even though it was the coldest day of the trip. The likeness we see in a mirror isn't more exact or more real than a painting, but it ages with us. My mirror still hangs between my childhood desk and my bed; that French portrait is in some closet, forgotten. In it, I'll always be fourteen. How old is the image of myself I hold inside? Maybe it's also fixed, and that's why my own face surprises me sometimes. *That is me.*

The artist began to be the subject of her own paintings in the fourteenth century, when mirrors had become more affordable. All throughout the month of December, alone in the Residencia, I think and write about mirrors; when I look into them, they invariably reflect me. I don't want this project to appear to be a self-portrait, but I can't ignore the fact that these thoughts have a body. They aren't just suspended in the ether.

I COULD BEGIN TODAY: January 5. Álvarez Gato alley is near the Plaza de Santa Ana. As we get closer, I realize I've passed by here many times. The Poet guides me to the entrance to the pedestrian alley. There's no sign of the nineteenth-century ironmonger's shop with its two curved mirrors hanging outside the door to attract clients. On both sides of the street, taverns are built one after the other, with apartments on the upper floors. Flanking the door of the bar Las Bravas hang replicas of the concave and convex mirrors of days gone by. They're smaller than I had imagined. We take ourselves in—the concave mirror squishes us, the convex one stretches us out. Ramón María del Valle-Inclán created an entire literary genre out of them: esperpento. The day before, I'd read the part in *Bohemian Lights* when Max Estrella talks with Don Latino about the mirrors in Álvarez Gato alley, calling them absurd. He wants to deform them, just as he and Don Latino are deformed in them. Max Estrella declares: "Spain is a grotesque deformation of European civilization," much as the reflection is of his likeness.

I tell the Poet what I've been writing: about the reflections I find around Madrid, and what it feels like to study myself in the mirror. But she tells me she experiences seeing herself differently. She never forgets her own image—it follows her wherever she goes, immense, overwhelming. Every glance in every mirror reveals the same defect, a disproportionate size she has learned isn't real. "What I see isn't what others see." She doesn't need curved mirrors, her brain already does the work of deforming her likeness so that it's grotesque to her. Every morning when she gets dressed, she follows the same routine I do: she picks out clothes she's excited to wear, eager to see herself as she is. But then she looks in the mirror, regards her body, and, despite the anxiety, forces herself to remember an admonition that has become a personal mantra: "What I see isn't me, because I don't know how to see myself." I can't tell her I don't understand, or that I instinctually think: *at least that doesn't happen to me.* I suppress the thought, grab her hand, and find it's hot.

No one fits between my reflection and me. But as we talk, I step to the side and let the Poet enter the text, and she returns my gaze.

I COULD BEGIN WITH an experiment published by the University of Liverpool. It asks us to imagine that we're standing at the bathroom mirror. Imagine it in detail. Our hands resting on the white sink. The blue-tiled wall behind us. Containers of various sizes, full of liquids and creams, are grouped together with the soaps and toothpaste, the landscape of a routine. Now that we have the scene in our minds, we have to hold it there while answering two questions: What size is the copy of your face in the mirror? What would happen to it if you started walking backward?

Most people answer that the size of their reflected face is equal to the size of their real face, and that as they back up, the image gets smaller.

Most people are wrong.

It's nothing more than a game of light and angles.

I COULD BEGIN BY putting the experiment to the test. With one eye closed, I trace the outline of my face on my closet mirror with a blue dry erase marker. It's impossibly small. I take a step back, and my face fits perfectly. One more step, another, another, but no matter how much I back up, my face stays its original size. Leaving the blue marks there, I go out into the hallway and knock on my neighbor's door. He's a mathematician. I explain what I'm doing, that I know it's counterintuitive. But for him, it makes all the sense in the world.

We repeat the experiment. Our faces maintain their size regardless of how far we are from the mirror.

I tell him I don't get it, and that if I don't understand it, I can't write about it. He asks me to imagine that what I see is just reflected light, and that the mirror is a flat, two-dimensional surface halfway between the real image and the virtual image. He tells me that the space around us has Euclidian geometry, but we see it in hyperbolic geometry. "Sounds beautiful," I say, "but tough to imagine." He explains that when train tracks disappear on the horizon, appearing to converge at a single point in infinity, that's an effect of hyperbolic geometry. Standing at his closet mirror, identical to mine, he tells me that we perceive angles and not sizes, that what we see in the mirror is something of an optical illusion, a mirage.

I COULD BEGIN TODAY: January 7. For my birthday, I buy myself a kaleidoscope because with all this writing about mirrors, I can't stop thinking of the one I had as a kid. I want to recreate the memory of twisting it to produce a collage of fragments, giving way to ever-new combinations as they shift.

These toys drove English society wild in the nineteenth century. Kaleidoscope factories couldn't keep up, and demand quickly outpaced supply. At first, they were just entertainment for adults. Now you can buy kits with all the supplies you need to build one yourself. You need a tube, such as a paper towel roll, three mirrors, and tiny colorful pieces of glass or plastic. The mirrors are placed inside the tube, forming a triangle. When one mirror is reflected in another, an infinite plane is created. But, if the mirrors are placed at an angle, the duplication becomes symmetrical, covering the entire surface. The duplication extends without creating any gaps or overlaps when the mirrors form an equilateral triangle. I spend a lot of time trying to spot the smallest unit, the original triangle out of which the others sprout, and although sometimes I think I've found it, other times it gets away from me.

Days later, I realize what I bought isn't a kaleidoscope, but a teleidoscope. Instead of colorful fragments, it has a magnifying glass, and whatever's outside it forms a picture that gets broken down and multiplied. I like pointing it toward lights or colorful objects. It's a decomposition scope, allowing me to twist and separate whatever surrounds me into geometric forms and colors, to zoom in or to gain distance. I think it's a good metaphor for what I'm writing. With each rotation, I search these images, these words, for new shapes and other ways of seeing.

I COULD BEGIN WITH the idea of the virtual. In physics, the virtual has an apparent existence but isn't real. That's why a reflection is a "specular or virtual image."

Specular.

An adjective used to describe symmetrical things, like an object and its double. Relating to mirrors. Or *speculate*, a verb meaning to make conjectures based on insufficient evidence, or to reflect on an exclusively theoretical plane. To *reflect* comes from *reflection*, as in thinking, or as in a visual representation, or that which is *reflexive*. All this language is like a game of mirrors, multiplying to infinity whatever it touches. Even this essay is becoming a reflection of the act of writing it, a Borgesian game of writing an essay about writing an essay, with the author at the center, multiplied in all the mirrors of her life at once, trying to figure out which is the real version.

This brings me back to *mirage*: "an illusory image." I wonder if the word might have come into existence not because of the optical illusions visible on the highway, but as a reminder that the world of the mirror is a world of unreal images, images that can never serve as anchors. Every attempt we make to see ourselves has failed before the first glance.

I COULD BEGIN IN literature. Narcissus drowned in his own reflection. Medusa is petrified when she catches sight of herself. The witch in "Snow White" needs someone to confirm her beauty. The mirrors of Galadriel and Erised reveal people's deepest desires. Alice travels to a parallel world. As a vampire, Dracula casts no reflection. Borges and Arredondo hate mirrors for their capacity to multiply. Ashbery writes his self-portrait in a convex mirror and Vilariño buys one for her bathroom. Rilke speaks to them directly, telling them no one has ever described their essence. Woolf believes people shouldn't hang them in bedrooms. They appear in the horror stories of Poe and Hoffman; in the poems of Plath, Hardy, and Lorca; in works like Shakespeare's *Richard II*. I could go on, but literature is so crammed full of mirror images and gazes that just a few are enough to create the feeling of an infinite mirror.

To say that "art is a mirror" is cliché. But what is it, if not Cortázar's axolotl? Sometimes I forget that writing is *my* form of reflection. Little by little, I start to accept that each new beginning of this essay is just one piece of the full picture. The anecdotes and facts in my notes interweave, reference each other, form clusters. Though I cling to each individual idea on its own, I'm so close to them that I can't see the picture they make up. I discern its outline as I write it.

To speak of mirrors and literature is to bite my own tail and return to the beginning.

I COULD BEGIN WITH any one of these pages, and I would always arrive at the same point. Today. Before going out, I open my closet door. A shock of unfamiliarity runs through me. I take in the person wearing my clothes, my glasses, my shoes. *Who is that?*

THE HISTORY OF SEEING

Object of Study: Light

IN THE BEGINNING there was light. In all the beginnings. In all creation stories, in philosophical debates, in every question we have about vision. In the dawn rays that slid over the walls of the cave, drawing us outside. Light always comes first. Tonight, when I get home, I turn on all the lights in my room out of pure instinct. Fear of darkness is our first fear. From fire to LEDs, history can be recounted through the measures we take to keep darkness in check—to understand light. But by now it has lost its mythic air. I close my blinds so not a photon can escape, trapping them between these four walls. For millennia light was evasive, incomprehensible, uncontrollable; its nature shrouded in shadow.

I use the words "turn on" or "light" when I'm talking about lamps, candles, screens. The flash of my cell phone camera. The sun. The brightness radiated by celestial bodies during combustion. A representation of God. Light bulbs: warm, cool, flickering, green, blue, dim. Sparklers, highways, intersections. A measure of galactic time. We use the concept of light to talk about birth, explanations, comprehension, clarifications, emphasis. We say "a light bulb went off in my head" because light symbolizes truth, goodness, and knowledge. We talk about the "light of my life," the light of reason, or the light of freedom. Defining light as the physical agent that makes objects visible is too naive. Better to think of it as the go-between for energy and matter. Or as José Lezama Lima put it: "Light is the first visible animal of the invisible."

Euclid was the first to mathematize light, claiming that the eye emits a ray in a straight line. He steered clear of words, explaining it instead through angles, rays indicating direction, and collision points. Fifty-eight propositions, each followed by a proof that concludes: "And that is what we wished to demonstrate." It took me a long time to accept that writing helps me understand the questions, rather than nailing down the answers. But sometimes I long for the clarity and security of math, to be able to simply state, "this is what I wished to demonstrate." The danger lies in basing, as Euclid did, one's work on an error of interpretation. His geometry assumed that light travels from the "interior fire" of the eyes toward objects. And yet his mathematics persist. The angle of incidence of light is equal to the angle of reflection. Given two points, a straight line can be drawn between them and parallel lines never intersect. Observations-turned-geometry marked the beginning of optics.

In early March, I get a text from the Physicist with a quote from Niels Bohr: "Science, like poetry, does not describe facts but creates images." I try to track down the original source, but can only find a conversation in Werner Heisenberg's memoir. It begins with the question of how to understand atoms if there is no language to describe them. Bohr answers: "When it comes to atoms, language can be used only as in poetry." The poet is not concerned with describing facts, but with creating images and forming connections in the mind. I respond to the Physicist with: "See, Bohr proves me right" along with the quote, which ends: "I think we will reach an understanding of the atom, but in the process we may have to learn a new meaning of the word *understanding*."

Aristotle said that light was an "activity—the activity of what is transparent." He said that it could not travel from eyes to objects because, if that were the case, we'd be able to see in the dark. And that if it traveled from objects to eyes, the problem lay in how it was propagated—if it propagated like sound, which displaces air by undulation, rays of light would have also to undulate some medium between the observer and the object. A vacuum wasn't possible; everything appearing to be empty space must be occupied by a thick, perfect, transparent substance, the fifth element of nature. He called it *ether*, for the Greek god of light, luminosity, and pure air. He said it filled the sky and all the spaces between the stars.

In Latin there are two types of light: direct/*lux*, and reflected/*lumen*. These words are now units of the international system. Lumen (lm) measures the total amount of visible light emitted by a source. Lux (lx) is the measure of illuminance, that is, lm/m^2, or the incidence of light on a surface. One lux is equal to the measure of the light of the full moon at zenith in the tropics, while the darkest limit of twilight beneath a cloudless sky is equal to three lux. The cloudy night sky during a new moon measures 100 μlx, and the light of the sun at midday is 100 klx. I think about the illuminance of the sky in Mexico City and Madrid. It's measurable—if you wanted to, you could assign an exact number to the difference. I think about how lumen and lux are defined through candela (cd), which measures luminous intensity and is one of the seven basic units of the international system. Although candela is now defined based on the second and the watt, the old definition is still valid: 1 cd is equal to the brightness of a candle in a dark room.

The Middle Ages are sometimes called the Dark Ages. Light divides humanity into those captivated by its divine qualities, and those who must measure and explain it. In Europe, this inquiry withdrew into the monasteries, where mathematical beauty was studied, and light became "the beauty and adornment of visible creation." Gothic cathedrals reached pompously toward the heavens to capture the divine splendor and luminosity. Light, a mixture of theology and mathematics, shone in through the stained glass and the rose windows. It bounced off buttresses and arches like a metaphor for God.

In planning a third essay, I think I'm going to write about the eye, but the first thing I do is read the history of optics. I mark entire paragraphs, copy down facts, and stay up till all hours listing every discovery and the connections between them. I'm obsessed. Little by little, I begin to make out the scaffolding of a text bringing humanity's first question, "What is it to see?," into my present moment. I don't understand why I'm feeling so moved, but I can't curb my curiosity. I have to write my own version, even if it means pulling away from myself and closer to science.

During a partial eclipse of the sun, light filters between the leaves of the trees I walk beneath, and the ground is covered in half-moons. Aristotle witnessed this phenomenon and wondered why light maintains the outline of its source when passing through irregularly shaped openings. Through the square holes in a basket, for example, beams of light form circles. But the question went unanswered until the turn of the first millennium.

To survive the Middle Ages, optics paved the way for the sciences to travel from Alexandria to Baghdad, from the burning library to the House of Wisdom. There, Ibn al-Haytham was the first to use a camera obscura for research. He allowed light to enter through a tiny hole in the wall of a windowless room—but the inverted images, which would one day give rise to photography, didn't interest him. He followed the path of the light, measured its trajectory, and proved that light rays didn't mix but instead were straight and independent. Ibn al-Haytham observed an eclipse reproduced hundreds of times on the floor—and thanks to the camera obscura, answered the question that had intrigued Aristotle. The apertures between the leaves allow light to pass through, recreating the shape of the emitting source.

"The art and science of asking questions is the source of all knowledge."

One night in October 1604, a new star appeared in the sky. It shone by day and night and was visible to the naked eye for eighteen months. It was not a star at all, but the death of a supernova inside the Milky Way.

Galileo observed it from Padua, Kepler from Prague. With their telescopes, they both measured the universe. Their observations demonstrated that the firmament was not static: the earth, the stars, light, everything was in motion.

Kepler believed that light's velocity was infinite, but Galileo was not convinced. What was it made of? If it was made of particles, their velocity had to be finite. Galileo proposed an experiment in his *Dialogues*. Sagredo and Salviati would climb two mountains a kilometer apart, each carrying a lantern. Sagredo's would be lit, while Salviati would light his upon reaching the summit. As soon as Sagredo saw the light, he was supposed to swing his lantern, and Salviati would count the time the light took to travel between the two mountains. Though Galileo never carried out this experiment, using this metaphor he was the first to suggest that "if light is not instantaneous, it is extraordinarily rapid." Before his death, under house arrest and suffering from glaucoma, he confessed that he would have spent his life in a cell if he could have understood what light is by the end.

As I read and write the history of light, I try to remember my aims. It's easy to become totally engrossed in the facts, so I return to the question "how can I make poetry out of science?" I set guidelines for myself: concentrate on images, be unfaithful to scientific meaning, always look for the metaphor. When I get lost in my experiments, I pull myself back to the definition of metaphor: one element used in place of another to suggest an analogy, such that a new meaning is created. It's a tool for talking about unnamed things. It's a rhetorical device, a kind of figurative language to conjure up reality but disguise it at the same time. It transfers the name of one thing to another, according to Aristotle. It attempts to contain not only the concept but also our subjective perception of it. Ortega y Gasset called each metaphor the discovery of a law of the universe.

Leonardo da Vinci described birds as instruments working according to mathematical law. He believed in studying the science of art and the art of science. His paintings, his advice to other artists, his exploration of perspective, his anatomical studies, and his flying machines all arose from this idea. In his treatise on painting, he advised his acolytes to look at light and consider its beauty, its effect on the landscape. I envy the way da Vinci achieved making something scientific out of the artistic, just as I want to make something personal out of the scientific. I scan one of his paintings and there, in the background of *The Madonna of the Yarnwinder*, the mountains look bluer and more diffuse the further away they are. Da Vinci studied Ibn al-Haytham, learned about perspective, and applied the laws of optics—"the blood of physics"—to evolve art.

It's surprising to me that it almost never rains in Madrid. I don't realize this until the first storm. I open the window and the sound of it enters my room. It comes and goes, doesn't even last ten minutes. It leaves the garden smelling wet and a rainbow on the horizon. All I can remember is that the water suspended in the clouds splits the sunbeams, separating light into all its colors. Who has ever explained rainbows? Aren't they a subject as artistic as they are scientific?

Staring into his fireplace, Descartes noticed particles engulfed in chaotic, violent motion. When the flames so much as quivered, he could feel the light. He concluded that this was the pressure of the ether on the eye: pressure that could be bounced, bent, or blocked, and could propagate. It traveled in a straight line and radiated out in all directions from luminous bodies. That mystery solved, Descartes decided to tackle explaining rainbows, so he came up with an experiment. He stood at his window with the sun at his back and held up a fishbowl like a raindrop, allowing him to measure the refractive index of water for the first time. He moved the fishbowl up and down and found that a rainbow formed if the angle between the incoming and outgoing beams was 42°. That's why rainbows never appear at midday and are always seen at the same height above the horizon: as you walk toward one looking for its end, the colors recede, the angle stays the same.

Descartes also believed that light travels through water faster than air. Fermat didn't agree. When a lifeguard is rescuing someone from the ocean, they run along the shore as far as possible before entering the water. In the same way, light—which always takes the shortest path between two points—also covers more distance through the air. But it's not Descartes and Fermat who bring the idea of refraction to mind for me, but rather the Mathematician. I think of the Cartesian axes and Fermat's Last Theorem. I think of the evenings I spent sitting at the kitchen table while she helped me with my math homework, saying, "let's do this problem too, because it's very beautiful," even if it wasn't assigned as homework. I remember how she taught me matrices to solve simple equations, and how when I was six years old, just learning to write out my numbers, she started with the negatives. The beauty of mathematics made my mother want to show me things I wouldn't understand until many years later. I also think of the mathematicians she told me stories about. Like Évariste Galois, the first to use the word "group" in a mathematical context, who died in a duel at the age of twenty. The night before he died, he had frantically written letters to his friends containing theories that revolutionized mathematics. Or Sofya Kovalevskaya, the first woman to earn a PhD in mathematics in 1874, who was also the first woman to hold a position as a university professor in Europe. "It is impossible to be a mathematician without being a poet in soul," she wrote. When I come across this quote, I send it to my mother. Then I send her another from Ada Lovelace: "Mathematical science shows what it is. It is the language of the unseen relations between things. But to use and apply that language we must be able fully to appreciate, to feel, to seize, the unseen, the unconscious." The next day, my mother answers that mathematics is not a science but a language, and to dedicate your life to it is like dedicating yourself to art.

After my first semester of university, once I'd filled out my course evaluations, I responded to a socioeconomic survey that asked: "How many light bulbs are there in your house?" I went through every room, counting. "More than ninety," I answered. At the time, the question surprised me, but now I understand it. Light has always been a symbol of status and power. Before light bulbs, it was candles, fireplaces, oil lamps, bonfires. Before public lighting, ostentation remained inside the home. Lights reached the streets for the first time in 1666, when Louis XIV decreed that hundreds of lanterns be hung along the boulevards of Paris. One after another, the European capitals copied him, conquering the monsters of darkness, life seizing hold of the night. When I cross Madrid at four in the morning, I take special pleasure in walking the deserted streets. They don't make me nervous like they would in Mexico. I don't imagine danger lurking in the shadows. Sometimes I pause to look at the street-lamps and think about how we take the light around us for granted. Most of human history was carried out in shadow, but after years of tests and measurements, light was set free—and with it, humanity.

Scientists of his era would have said that Isaac Newton was cagey and conceited but unquestionably a genius. He described himself as a child playing in the sand at the seashore, finding a smoother pebble than the rest—or, as a wave receded, a more beautiful shell. The mysterious ocean of truth licked the sand just out of his reach, waiting to be discovered. Isaac Newton's curiosity was powerful. Once, he looked at the sun's reflection in a mirror and was blinded for three days. He stuck a blunt needle between his eye and its socket to study how we perceive colors: when he applied pressure, he observed dark and colorful circles that disappeared little by little. When he ran into problems describing what he saw, he invented calculus. He defined force, mass, movement, and acceleration. He reproduced the experiential world using mathematics, describing it and putting it into words. When he tackled light and the dilemma of color, he said that the answer "depended on imagination and fantasy and invention." Maybe Newton could have been a novelist, because as it turns out, he never went to the ocean and barely ever left his hometown. But from there, he was able to unravel the laws of nature.

An "experimentum crucis" is an experiment capable of determining conclusively whether a theory is superior to others, given the same specified conditions. Newton once bought two prisms at a market, and, since the plague was devastating England, locked himself in his room. Through a hole in the window shade, he sent a beam of white light toward one of the prisms. On the opposite wall, a rainbow appeared. As the colors passed through the second prism, the resulting light had no color. White light, he concluded, is formed by mixing all the colors. Then Newton blocked all the other beams of light except red—this was the crucial step, revolutionizing optics forever after. The thread of red light pierced the second prism, and the result was red. He repeated the experiment with orange, green, blue. He got only orange, green, and blue. Each color passed through the prism and came out intact on the other side, pure and indivisible. White light, and not the object, was the source of color. The green of the grass, the blue of the sea, the yellow of the sun are not intrinsic characteristics: what we perceive with our senses, what we take as truth, is no more than the process of the absorption and reflection of light.

Newton's physics are called "classical physics." They govern the world on the scale at which we encounter it, neither macro nor micro. Our experience is Newtonian. Our language is Newtonian. Let's imagine, for example, a bubble floating in the sunlight. As it floats, swirls of rainbows form on its surface. Feynman said this phenomenon is impossible to explain in a Newtonian sense. Attempting to do so puts you at the heart of quantum mechanics, takes you straight to its greatest mystery. But even so, once a phenomenon is understood, an attempt should be made to express it in classical terms. Bohr defined "experiment" as a scenario in which it's possible to recount to others what you've done and learned. "The account of the experimental arrangement and of the results of the observations must be expressed in unambiguous language with suitable application of the terminology of classical physics." Therein lies the problem.

Goethe heard about Newton's experiment and bought a prism. Instead of observing how light entered it, he looked through it at the window mullion. In other words, he looked at a dark line against a lit-up background—the opposite of Newton's experiment. He didn't see red, green, and blue, but magenta, cyan, and yellow. The poet concluded that color originates when white penetrates black, and thus light isn't the synthesis of all the colors but their origin. Yellow is light diffused by darkness. Blue is darkness weakened by light. In 1810, Goethe wrote *Zur Farbenlehre*, or *Theory of Colours*, in which he describes his experiments and conclusions. For him, color is not a property, but a personal, subjective perception. We know the composition of light by the effects it produces. Yellow is happy, serene in nature. Red is serious and dignified. Blue is moody, the color of what's far away, like the mountains and the sky. Maybe this is the inflection point I'm searching for between science and art. Maybe the debate between properties and perceptions, between objective and subjective, is whatever sets them in opposition. Maybe the split happened precisely when Goethe declared that Newton's mistake was trusting mathematics more than what he saw with his own eyes. Maybe I should ask myself whether I'm not making the same mistake as a writer.

The story goes that the British Romantic artists gathered for a dinner party, and Keats raised his glass to make a toast: "To Newton's health and the confusion of mathematics, which have reduced the beauty of the rainbow to the size of a prism." From Chopin to Keats, the Romantics worked in pursuit of the sublime, discounting the analytic and rational, such as Newton's empiricism or Kant's pure reason. Light was the essence of the transcendent and beautiful. They preferred Goethe's theories, which saw and sensed light, over those of Newton, who had reduced it to numbers. Is it here where the scientist's cold gaze was invented, that look incapable of marveling at nature? Humans see the moon and wonder what it's doing up there, what it is, why its light is colder than the sun's. Is this how art originates, or science? Both describe the world and experience through observations. One does it using objective language, the other using subjective. One is based on perceptions, the other on properties. One attempts to predict, the other empathizes. Two faces of the same coin. And what do I do? Observe science, write about it, try to reclaim the sublime. Goethe also admired science. He carried out each of his experiments with the precision of the scientific method. They followed the rules: empirical, repeatable, and systematic. Like his scientific peers, Goethe was obsessed with explaining what he saw. Maybe that's why legend has it that, on his deathbed, his last request was "mehr Licht." Open the windows and let in more light.

"Poets say science takes away from the beauty of the stars—mere globs of gas atoms. I too can see the stars on a desert night, and feel them. But do I see less or more? The vastness of the heavens stretches my imagination—stuck on this carousel my little eye can catch one-million-year-old light. A vast pattern—of which I am a part. What is the pattern, or the meaning, or the why? It does not do harm to the mystery to know a little more about it."

Theodore Brown, the author of the general chemistry textbook I used in university, was fascinated by metaphors and their effect on the scientific process. For him, observations, theories, and the models we use to explain a given phenomenon form a relationship. Metaphors are located at the center of this relationship because they allow a new or abstract concept, something we cannot experience, to be expressed through something known. I have to admit I rarely use metaphor in my own writing.

Maybe that's why I hadn't realized how often they're used in science until now. Light is like billiard balls, Newton said, that's why it bounces. Light is like ocean waves, said Huygens, that's why it refracts, polarizes. "Scientists make up stories about how the world works," writes Brown. These stories are based on what they perceive through their senses or their instruments, and then they put these perceptions to the test through modeling and experimentation. Stories, models, and theories are tested and refined through our experience of the world. Light is a wave, a particle; it's both. Each metaphor carries with it one way of thinking about light. Explanations and experiments change: particles are matter, waves are movement. In this way, language splits the path of science in two.

When someone asks what I'm writing, I take it as an invitation to tell them about light, to talk through the historical moments I've selected. I don't trust my own judgment. I feel too close to the material, caught up in the excitement of it. Sometimes I want to send these sections to my friends—either writers or scientists, it doesn't matter. Sometimes I stop myself, but other times I can't hold back. I ask them, "Does this make sense?"

when what I really want to know is whether they find it as fascinating as I do. I crave their reassurance, because I feel as though I'm juggling a bunch of tiny crystal figurines, trying to get them to move in sync. I'm afraid I'll trip over all the fragments I've collected before I get to the end of this story, and I'll wind up here in my room with all the lights on, not having understood a single thing about light or about myself.

The eighteenth century was the century of light: of the Enlightenment, the end of thinking in darkness, the beginning of the social and scientific revolutions. Still, for me, light really took the stage in Paris during the nineteenth century. Discoveries overlapped and conflicted. Some canceled each other out, but most began to harmonize, gaining strength until they unleashed a scientific and artistic revolution. Daguerre took the first photograph, the Lumière brothers invented the cinema, Edison brought electric light into homes, Faraday polarized a beam of light with a magnetic field, Monet painted the Rouen Cathedral more than thirty times, Maxwell described the laws of electromagnetism, Thomson discovered the electron, a yellow line observed in the solar spectrum during an eclipse led to the discovery of helium. Everything is unleashed, is accelerated, is illuminated.

In 1803, Newton was believed to have been correct when he said that light was a particle. Then Young thought up an experiment to test its wavelike characteristics. He cut a small hole in his window shade and covered it with a piece of paper perforated with a pin. With a mirror, he deflected the thin thread of light. Then he placed a 0.2 mm-thick card in the path of the beam: it split the light in two. As if two rocks had been thrown into a lake, the result projected on his wall was a pattern of interference. One line of light, a line of shadow, a line of light, another of shadow. The bright lines were created when two crests of a wave became superimposed, reinforcing each other. The dark lines were the effect of the crest and a valley canceling each other out. This phenomenon could only be explained if light acted like a wave, given that it was impossible for a particle to be in two places at the same time. It's another experimentum crucis. Once again, our understanding of light changed from particle to wave.

Metaphor is tinged with ambiguity. We understand and create it intuitively, almost automatically. It clashes with the idea of scientific thinking as logical and rational. But science is not independent from the human. Just like everything human, science is suspended in language and in its own ambiguity. The challenge lies in accepting its uncertainty, embracing it in what I write and not running away from it. For example, I think about resonance, the propagation of sound. It's the phenomenon produced when the frequency of the system under study coincides with the frequency of an external system. They vibrate with the same rhythm. But isn't there also a resonance when I come across Gabriel Dawe's artwork? Like me, he's a Mexican obsessed with light. I see a photograph online of one of his installations, made of threads. It looks like a static rainbow, just brushing the roof. I observe a close-up; I'm impressed by the precision it must take to pick out each thread. His installations imitate the decomposition of beams of light: purple, pink, blue, multicolored. A game of threads in patterns that seem to filter in through a window or radiate out from a light bulb, propagating through the space. They're called *Plexus*, like the network of blood vessels and veins that runs through the body, making up the nervous system. The name is also a resonance—from science to art and back to science. As I write, I get the feeling that everything I live out, read, or come across is interlaced, creating a kind of woven tissue, almost an anastomosis beating between my notes and the text. I hope that the end result resembles one of Dawe's dazzling cascades.

I can't remember when I first studied the electromagnetic spectrum. I was probably in high school when I saw a diagram of the frequency bands accompanied by a sine wave tightening toward the right. The wavelength and energy were always written below. Radio waves, the largest waves with the slowest vibrations, were to the left; gamma rays, small waves that vibrate rapidly, to the right. The 300 nm of the visible spectrum at about the halfway point, with the colors of the rainbow from red to violet. Until 1800, this was all the light that existed, but little by little infrared rays were discovered, then radio waves, X-rays, microwaves, and ultraviolet light. No one knew that visible light and these new radiations were connected until Faraday realized that light responded to magnetic fields. Then came James Maxwell. It catches my attention that he also wrote poetry, and I look up his poems. Many of them are about his fascination with science, describing areas of study and scientific instruments. I take special note of one title: "Lines written under the conviction that it is not wise to read Mathematics in November after one's fire is out." Maxwell proposed the four equations that synthesize the behavior of electromagnetic waves. This is how the electromagnetic spectrum is constructed: it's engineered. We heat up our food. We communicate by transmitting and receiving waves. We transform and study matter. Light gets redefined and is no longer what we perceive with our senses. It becomes invisible.

Writing about the electromagnetic spectrum, I can't help but think of Marie Curie, about the radiation of alpha, beta, and gamma particles, about radium. "I am among those who think that science has great beauty. A scientist is like a child placed before natural phenomena which impress him like a fairy tale," she said. Radium shines with a pale blue light. It's luminescent. I think about the clocks whose faces and numbers glowed in the dark because they were covered in radium paint, about the Radium Girls who were in charge of painting them in the 1920s. They died of cancer because they touched the paintbrushes to their lips, ingesting radioactive paint. I think about other types of luminescence that don't kill but occur in nature: in certain kinds of mold and jelly-fish, and in camouflage or communication methods. About photographs of algae lighting up the ocean at night. I think about the rabbit named Alba, "a living being and work of art at the same time," who seems like something from a fairy tale. In the image the artist presented, anyway, this rabbit injected with jellyfish DNA shines a bright green. But art, unlike science, tends to bend reality, and it's impossible to know whether the photograph is true or false.

We speak of electricity as if it were water: it has a current, flow, and density. Although the analogy with liquids is only used as a teaching method, at some point it was a serious theory. Benjamin Franklin proposed that electricity and heat were fluids. This model could not explain many behaviors, among them the repulsion of charges, electric fields, or conductors. Metaphors can hamper science when they become technical terms and lose their initial metaphorical significance. Or when we take them literally, forgetting that every metaphor is an approximation. Or when metaphors confuse the model with the modeled. Eventually, we found another way to conceptualize electricity, and Franklin's theory was dismantled—if not its vocabulary. Electricity travels through power lines like water through a river, but instead of roaring, the current produces a buzz. It illuminates.

Whenever I call home, I spend a great deal of time telling my parents the history of light, despite the fact that they both know it already. I jump from one era to another, from one scientist to the next. The Physicist asks if I'll write about ether. The Mathematician wants to know if I will delve into the importance of her discipline. Then I tell them about language, the metaphors I'm finding, how science uses them, how I want to use them myself. Right after one of our calls, I think of the perfect example. Recently, the Physicist has been using a computer program to study the arrangement of molecular patterns under different forces. He's been searching for the conditions necessary to form a specific pattern called the "Kagome lattice." In his program, little white balls arrange themselves, bump into each other, readjust. And finally, one day, *there it is*. On the screen, he tells me, the little balls arrange themselves like a Japanese bamboo basket.

At the end of the nineteenth century, no one had seen this luminiferous (or "light-bearing") ether. There was no proof of its existence, but if light was really a wave, then it needed a medium in order to propagate. Just as sound distorted air, light must do the same with ether. And so, following this metaphorical conclusion, ether was theorized and experiments were carried out. Scientists searched for the "ether wind" the earth must make as it moved through space, and which must block the rays of the sun. On July 8, 1887, Albert Michelson and Edward Morley carried out the most famous failed experiment in history. In a basement at sea level, they built an interferometer. This apparatus of mirrors and prisms, which separated and reflected beams of white light, floated in a trough of mercury. With it, they measured the velocity of sunlight. Time and again, they measured it. Time and again, they were incapable of detecting the ether. Time and again, they refused to accept that the universe was empty and that electromagnetic waves could not propagate, that one hundred years of theories were incorrect, that the problem was not the experiment but rather the idea behind it. Finally, they had to accept that giving a name and properties to a substance doesn't make it real.

How do you name something that has never been named before? Language can't encompass it completely, not all observations can be expressed mathematically, and the scientific process isn't linear. Neither is my sense of how to put a book together, how to visualize the end result and close the gap between what's known and unknown. The paradox of scientific metaphor is that it's necessary but invisible. Scientists forget about it. Miroslav Holub said that the words in scientific articles do not aspire to be assertions in themselves, but rather they exist for the verification of future experimentation or of a given theory. When a word or expression is defined in mathematics or physics, it doesn't mean anything beyond its definition. But science can't only be objective abstraction, just as writing can't only be about the subjective and sentimental. There's a place, possibly unnamed, where they coexist.

I find an image of a lit match online. The shadow of its wooden stick is projected onto a wall, with no sign of the flame. Below the image is the caption: "How old were you when you found out fire has no shadow?" I read and reread this question, thinking: "Today, the age I am now." The more I consider this idea, the more obvious it seems. Fire, the flame in the picture, is an expression of the energy liberated during combustion, in the form of light and heat. I get a lighter to test it myself. In my dark room, I strike it and observe the shapes projected onto my wall by the flame: my hand. The rectangle of the lighter. A small gas halo. No flame. Light creates shadows, it doesn't cast them. Though it seems the most natural conclusion, it leaves me feeling unsettled.

Einstein said that when he was sixteen, he pictured himself chasing after a beam of light. He wondered what would happen if he could run in the same direction as the wave, so fast he managed to reach it, mount it, ride on it, hurtling through space together at velocity c. What would he see? The static electromagnetic field? That was impossible: it violated Maxwell's laws, and the laws of nature cannot be dependent on velocity. This thought experiment was the seed that would one day prompt him to propose the theory of relativity as he watched a streetcar go by. $E = mc^2$, one of the most famous equations of the twentieth century. The product is mathematical; it has a logical, abstract structure. But earlier, at the heart of this discovery, was imagination and scientific ingenuity. De Broglie said: "Science is ultimately a matter of feeling, or rather, of desire—the desire to know or the desire to realize."

In university, I studied all the atomic models. For centuries, it was accepted that the atom was the smallest unit of matter, indivisible and stable—until J. J. Thomson discovered the electron. Neither atoms nor electrons can be seen with the naked eye. We have to come up with metaphors to explain our observations. Thomson thought the atom must be like a plum pudding with the electrons uniformly dotted throughout, suspended in a cloud of positive charge. Then came Ernest Rutherford's model, the most common representation: the electrons orbit around a nucleus, forming a pattern somewhere between a star and a flower. His student, Bohr, proposed a model that looks like a small planetary system. These models were able to explain many properties, but not the consequences of quantum mechanics. Subatomic systems ceased adapting to the metaphors of our macroscopic experience, from the planetary model of the atom to Sommerfeld's generalization; to Schrödinger's wave distributions, called "orbitals"; to Dirac's equations. I remember drawing each model over and over again to help me understand what they meant, what they predicted, and how to use the knowledge they gave me. In order to do that, you have to think of each one as a construct. They're flexible metaphors for reality that use correlations, predictions, and interpretations of results to continuously create new experiments. Now I work with different kinds of constructs: stories, essays, poems. I also try to use them to capture something of the macroscopic experience. My own representations start with my observations and go through my imagination onto paper, as I try to capture sensations as evasive as subatomic traces. In the end, all these models are the only way we can visualize a world beyond our measure.

A friend texts me a link. He knows I'm writing about light, and sometimes he sends me interesting facts. The link comes with no message or explication, so I'm not prepared for what I'm about to see. It's an article announcing the winner of a scientific photography contest, a PhD student at Oxford who called his work *Single Atom in an Ion Trap*. Throughout the next week, I keep coming back to this photograph. A cation of strontium floats in the space between a pair of needles set two millimeters apart. In order to capture it with a camera, the atom is bombarded with a violet-colored laser. A shining little dot, bluish, floats between the two massive points of metal, almost like a mote of dust. This is visual proof of the world beyond our senses. I wonder if we'll soon be able to observe the interior universe of that *indivisible*.

Every object emits energy. For example, a sheet of just-baked cookies. The wave model couldn't explain how they emit electromagnetic radiation at a certain temperature. So Planck quantized matter and subverted classical physics. To flesh out the math, he introduced a constant. The physical explanation is that radiation exists not arbitrarily, but rather in packets, called "quanta." Einstein later used this idea to figure out the photoelectric effect, proposing that objects emit not waves, but quanta of light called "photons." Small massless particles, capable of transmitting energy. Light once again became a particle.

When I write, I often wish words were more precise, more solid; I wish I could begin by stating the definitions, like in mathematics. Other times I like nuance, layers of meaning, the fluidity of figurative language. The duality of light reminds me of the duality of words. It makes me think about my own duality, too. A while ago I decided I didn't want to be a scientist, but sometimes I don't feel I relate to the world like a writer. If I had stayed a chemist, I wouldn't be so conflicted. Still, scientists must also translate math into words: from models to language, from observations to language. Translating math can create paradoxes. An example: x = "this sentence has six words." If x is false, then "this sentence doesn't have six words." See, such a simple paradox. How truly mathematical is the logic of language? I pause and remember that to write is to invent, and to defy that logic time and time again.

When you begin to study quantum mechanics, science also opens itself up to paradoxes: electrons choose all possible paths, waves are particles, and cats are both dead and alive. Nature is "absurd from the point of view of common sense." Quantum mechanics is a mathematical description, and math works. It describes, predicts, explains; it's consistent with experimental evidence. It becomes problematic only when you understand the result, when you must state it in terms of classical physics. Wave-particle duality, the duality shared by light and matter, is only apparent through the limitations of our language. How can the words we came up with to talk about what we perceive possibly describe something *beyond* our senses, especially in those instances when even what we *can* perceive slips away from us through the cracks in language? In the quantum field, we're incapable of knowing an entire system. These limitations have led to various interpretations of the wave equation, as well as of the principle of uncertainty. Science and philosophy rub up against each other. What does it mean to know? What does it mean that the wave function collapses when it's observed? What does it mean to make observations in physics, if we're not seeing the subject with our eyes? Is that the same as measuring? "Shut up and calculate," some say. They want to ignore the fact that right at the limits of our experience, science tells us a story beyond what we are capable of expressing.

In recent decades, the idea of "light pollution" has become popularized. In Mexico City, as in many capitals, we don't grow up beneath a starry sky. The view that left me breathless as a child was not the Milky Way, but coming around the mountain on the way home from Cuernavaca and seeing the stain of light from the city extending across the horizon like a kind of halo. Astronomers warn that this pollution hinders our ability to view celestial objects, while scientists keep looking toward the heavens. They're searching further out for answers, at the ends of the universe. Cosmic background radiation, the earliest light, is the primal source from which all others spring. We follow the traces of heat of these microwaves toward the past, toward the beginning, toward the origin.

In 2002, *Physics World* conducted a poll to select the most beautiful physics experiment in history. They announced nine finalists and a winner. Eratosthenes' measurement of the circumference of the earth. Newton's prism. Galileo dropping different objects off the Tower of Pisa or rolling them down ramps. Young's double slit. Cavendish's torsion balance. Foucault's pendulum. Millikan's oil-drop experiment. Rutherford's gold-leaf experiment. But out of all these, the chosen experiment was the double slit as applied to electrons. Until 1961, the idea of sending a single electron through a double slit was a thought experiment. Jönsson was the first to observe the interference of an electron, proving the duality that quantum mechanics had predicted for all particles. But, unlike the other names on the list, history has forgotten him. In the 1960s, the result didn't surprise anyone, and it's ironic to me that the most beautiful experiment was performed without any fanfare, because it confirmed something that was already suspected and assumed. Where is the beauty in an experiment? In the science that explains it? In its technical difficulty? I come up against the issue of judging science based on aesthetics. Should I go by what I understand about art, or what I know about science? Something tells me beauty is not in the difficulty or the novelty, but in the idea behind it, in what the experiment reveals about the world—and in this, it is similar to art. Like a painting or a poem, an experiment is a representation of reality that astonishes us; moreover, it gives us conclusive proof of what we already took for truth. The double slit applied to electrons is the observable proof of an incredible phenomenon. Electrons, just like photons, exist somewhere between a wave and a particle, beyond our language, in the quantum world. Deep down, matter itself faces dualities and interferences.

The word "metaphor" comes from "transference" in Greek. Transference of meaning. And what is a transfer, if not connecting two separate places by means of movement? Some metaphors fail and, instead of drawing two meanings together, push them further apart. "I feel trapped by its brightness" breaks down, because light represents freedom and openness in a metaphorical sense. Other metaphors become cliché: "your eyes are bright stars" forges no new connections. But still others, the ones that last, the ones that filter into our everyday language without losing their power, are like bridges: between experiences, between imaginaries, between disciplines. "To see the light, to bring to light, to shed light on" are more than metaphors—they exist because language tends to be insufficient. In spite of its limitations, we experiment, measure, write. It's what moves us to innovate.

"Light is the one thing humans can see with their own eyes but cannot touch. It does not have volume or mass, except at the quantum level, and cannot hurt you except in certain cases. Light is the fastest thing, but occupies no physical space; it is almost free. That's why it has always seemed magical to us."

I admit that this essay is not exactly about light. Neither is it about scientific language, and even less about metaphor. It's really about my own experimentum crucis.

It's about pushing boundaries. Starting a conversation with all the parts of myself; embracing my dualities. Studying words until I can inhabit both worlds as they do. Learning, once again, how to write about the sublime, about the familiar, about the origin.

In Madrid, the time changes at the end of March, and they cut down the trees outside my window. A thousand lux enter unfiltered. I close my eyes. I note the reddish orange on my eyelids. What else can I say? Spring arrives and the days get longer. Here we are again, facing the sun. After four thousand years and the obsession of countless artists and scientists, we're left with this warmth on our skin and all our uncertainty. We go on without knowing what light is.

SOURCES

The form of each essay, as much as its content, was influenced by multiple sources. In many cases, I consulted Wikipedia and the Royal Spanish Academy (RAE)'s *Dictionary of the Spanish Language* to give myself a preliminary idea of various concepts. What follows is a list of the most important books and websites:

Alberge, Dalya. 2007. "Welcome to Tate Modern's Floor Show—It's 167m Long and Is Called Shibboleth." *The Times*, October 9, 2007. https://www.thetimes.co.uk/article/welcome-to-tate-moderns-floor-show-its-167m-long-and-is-called-shibboleth-mwktzwgwbdb.

ASTM. "D4359–90(2019), Standard Test Method for Determining Whether a Material Is a Liquid or a Solid." Last updated July 15, 2019.

Barrett, Cyril and Tatarkiewicz, Władysław. *History of Aesthetics*. Vol. II, *Medieval Aesthetics*. Lavergne: De Gruyter Mouton, 2015.

Bök, Christian, *Crystallography*. Toronto: Coach House Press, 2003.

Borges, Jorge Luis. "Nathaniel Hawthorne." In *Other Inquisitions, 1937–1952*. Translated by Ruth L. C. Simms. Austin: University of Texas Press, 1964.

Brewster, David, Isaac Newton, and Richard S. Westfall. *Memoirs of the Life, Writings, and Discoveries of Sir Isaac Newton. Vol. 2*. New York: Johnson, 1965.

Brill, Robert. "Does Glass Flow?" Corning Museum of Glass, September 29, 2011.

Brown, Theodore L. *Making Truth: Metaphor in Science*. Champaign: University of Illinois Press, 2003.

Chang, Kenneth. 2008. "The Nature of Glass Remains Anything but Clear." *New York Times*, July 29, 2008. https://www.nytimes.com/2008/07/29/science/29glass.html.

Curie Labouisse, Eve. *Marie Curie: A Biography*. Translated by Vincent Sheean. Boston: Da Capo Press, 2001.

Curtin, Ciara. "Fact or Fiction?: Glass Is a (Supercooled) Liquid." *Scientific American*, February 22, 2007.

De Broglie, Louis. *New Perspectives in Physics*. Translated by A. J. Pomerans. Montréal: Minkowski Institute Press, 2021.

Díaz, Luciana. 2014. "La vida como obra de arte." *La Nación*.

December 21, 2014. https://www.lanacion.com.ar/lifestyle/la-vida-como-obra-de-arte-nid1753575/.

English, Kathryn, "Understanding Science: When Metaphors Become Terms," *Asp*, no. 19–22 (January 1998): 151–63, https://doi.org/10.4000/asp.2800.

Feynman, Richard. *QED: The Strange Theory of Light and Matter*, Princeton: Princeton University Press, 2006.

Feynman, Richard, Robert B. Leighton, and Matthew Sands. "Astronomy." In *The Feynman Lectures on Physics*, Vol. I, Section 3–4, 3–6. New York: Basic Books, 2011.

Flinn, Gallagher, "How Mirrors Work," in HowStuffWorks Science, August 5, 2009, science.howstuffworks.com/innovation/everyday-innovations/mirror.htm, consulted September 2, 2019.

Frezza, Giulia, and Elena Gagliasso, "Building Metaphors: Constitutive Narratives in Science," in *Metaphor in Communication, Science and Education*, edited by Francesca Ervas, Elisabetta Gola and Maria Grazia Rossi, 199–216. Berlin: De Gruyter, 2017.

García-Colín, Leopoldo and Rodríguez, Rosalía, *Líquidos exóticos*. Mexico City: Fondo de Cultura Económica, 1995.

Garrison, John S. *Object Lessons: Glass*. New York: Bloomsbury Academic, 2015.

Goethe, Johann Wolfgang. *Zur Farbenlehre*. Berlin: Berlin Hofenberg, 2016.

Haydon, B. and Alexander P. D. Penrose. *The Autobiography and Memoirs of Benjamin Robert Haydon, 1786–1846*. Whitefish: Kessinger Publishing, 2010.

Heisenberg, Werner. *Physics and Philosophy: The Revolution in Modern Science*. London: Allen & Unwin, 1959.

Heisenberg, Werner. "The Development of the Interpretation of the Quantum Theory." In *Niels Bohr and the Development of Physics: Essays Dedicated to Niels Bohr on the Occasion of His Seventieth Birthday*, edited by Wolfgang Pauli, 12–29. London: Pergamon, 1955.

Holub, Miroslav. "Poetry and Science." In *The Measured World: On Poetry and Science*, edited by Kurt Brown, 47–68. Athens: University of Georgia Press, 2001.

"Is Glass Liquid or Solid?," various authors, updated 2021, https://math.ucr.edu/home/baez/physics/General/Glass/glass.html.

Kovalevskaya, Sofya. *Sónya Kovalévsky: Her Recollections of Childhood*. New York: The Century Co., 1895.

Lacan, Jacques. "The Mirror Stage as Formative of the I Function as Revealed in Psychoanalytic Experience." In *Écrits: The First Complete Edition in English*, translated by Bruce Fink, 75–81. New York: W. W. Norton and Company, 2006.

León, Manuel de, and Ágata Timón. *Las matemáticas de la luz*. Madrid. Los libros de la Catarata, 2017.

Lezama Lima, José. "Las siete alegorías." In *Fragmentos a su imán*. La Habana: Ed. Letras Cubanas, 1993.

Toole, Betty A., ed. *Ada, the Enchantress of Numbers: A Selection from the Letters of Lord Byron's Daughter and Her Description of the First Computer*. 2nd ed. Mill Valley: Strawberry Press, 1998.

Maxwell, James Clerk. "Lines Written Under the Conviction That It Is Not Wise to Read Mathematics in November After One's Fire Is Out," Poetry Foundation, 2022. https://www.poetryfoundation. org/poems/45776/lines-written-under-the-convic-tion-that-it-is-not-wise-to-read-mathematics-in-november-after-ones-fire-is-out.

Mezger, Thomas G. "Instruments." In *The Rheology Handbook*. 4th ed. 291–348. Hannover, Germany: Vincentz Network, 2012.

Neumann, Florin, "Glass: Liquid or Solid—Science vs. an Urban Legend," in Internet Archive. Wayback Machine, 1996, https://web.archive.org/ web/20070409022023/http://dwb.unl.edu/Teacher/ NSF/C01/C01Links/www.ualberta.ca/~Bderksen/ Florin.html, consulted September 2, 2019.

Pauling, Linus. *General Chemistry: An Introduction to*

Descriptive Chemistry and Modern Chemical Theory. San Francisco: Dover Publication Inc., 2003.

Pendergrast, Mark, *Mirror Mirror: A History of the Human Love Affair with Reflection*. New York: Basic Books, 2004.

Robson, Shellie Jo. "The use of metaphor in scientific writing." MA diss., Iowa State University, 1985.

Rochat, Philippe, and Tricia Striano. "Who's in the Mirror? Self-Other Discrimination in Specular Images by Four- and Nine-Month-Old Infants." *Child Development* 73, no. 1 (January–February 2002): 35–46.

Ron, Antonio Martínez, *El ojo desnudo: si no lo ven, ¿cómo saben que está ahí? El fascinante viaje de la ciencia más allá de lo aparente*. Barcelona: Editorial Crítica, 2016.

Ruiz Mantilla, Jesús. 2017. "Doris Salcedo: 'Lo difícil es lograr una imagen invisible, una iconografía sutil.'" *El País*, October 3, 2017. https://elpais.com/cultura/2017/09/29/babelia/1506702218_231288.html.

Scholze, Horst and Michael J. Lakin. *Glass: Nature, Structure, and Properties*. New York: Springer-Verlag, 1991.

Schwartz, Gustavo Ariel and Víctor E. Bermúdez, eds. *#Nodos: una aventura intelectual que explora las fronteras entre los diferentes ámbitos del conocimiento*. Pamplona: Next Door Publishers, 2017.

"The Ninth Watch for the Ninth Pitch Drop," in The Tenth Watch for the Tenth Pitch Drop, accessed September 2, 2019. www.thetenthwatch.com.

Valle-Inclán, Ramón. *Luces de Bohemia*. Madrid: Espasa-Calpe, 1992.

Vilariño, Idea. "Cuando compre un espejo." *Poesía Completa*. Barcelona: Lumen, 2016.

Walmsley, Ian A. *Light: A Very Short Introduction*. Oxford: Oxford University Press, 2015.

Watson, Bruce. *Light: A Radiant History from Creation to the Quantum Age*. New York: Bloomsbury, 2016.

Zanotto, Edgar D. and John C. Mauro "The glassy state of matter: Its definition and ultimate fate." *Journal of Non-Crystalline Solids* 471 (September 1): 490–5. https://doi.org/10.1016/j.jnoncrysol.2017.05.019

OTHER LYRIC ESSAYS THAT SERVED AS INSPIRATION

Bonnaffons, Amy. "Bodies of Text: On the Lyric Essay." *The Essay Review*, 2016. http://theessayreview.org/bodies-of-text-on-the-lyric-essay/.

Buffam, Suzanne. *A Pillow Book*. Ann Arbor: Canarium Books, 2016.

D'Agata, John, and David Weiss. "Foreword." In *We Might as Well Call It the Lyric Essay: A Special Issue of Seneca*

Review, edited by John D'Agata and David Weiss. New York: Hobart and William Smith Colleges Press, 2014.

Hood, Dave. "Writing Creative Nonfiction: The Lyrical Essay." *Find Your Creative Muse* (blog), July 22, 2013. https://davehood59.wordpress.com/2013/07/22/writing-creative-nonfiction-the-lyrical-essay/.

Howells, William Dean. 2012. "The Lyrical Essay." *Los Angeles Review of Books*, March 18, 2012. https://lareviewofbooks.org/article/the-lyrical-essay/.

Manguso, Sarah. *300 Arguments*. London: Picador, 2018.

Nelson, Maggie. *Bluets*. Seattle: Wave Books, 2009.

Rankine, Claudia. *Don't Let Me Be Lonely: An American Lyric*. London: Penguin Books, 2017.

Shields, David. *Reality Hunger: A Manifesto*. New York: Vintage Books, 2011.

Singer, Margot. *Bending Genre: Essays on Creative Nonfiction*. London: Bloomsbury, 2014.

Zambreno, Kate. *Book of Mutter*. Los Angeles: Semiotext(e), 2017.

ACKNOWLEDGEMENTS

In October 2017, I received a grant from the City Council of Madrid to live in the historic Residencia de Estudiantes. Thanks to that grant I traveled Europe, lived in a wonderful city, made new friends, saw a lot of plays and movies, went to a lot of verbenas in city squares—and amidst the busyness, I somehow also found time to write this book. I would never have gotten the grant without Luis Muñoz's unconditional support, and I would not have written so much during my stay without the people who made a home for me at the Residencia de Estudiantes.

The Visible Unseen also came into being thanks to countless discussions, cafés, after-dinner conversations, emails, Skype calls, and walks with so many friends near and far that I can't begin to name them all. Thank you for lending me your words, for reading and rereading, for looking for typos, editing, giving me ideas—and above all, for listening to each one of my interesting facts. Among them, Kelsi Vanada and

Karen Villeda deserve special mention. The first for her translations and the second for her readings, for the title, and for entering this book, and others, into contests.

To the team at the Tierra Adentro Cultural Program: many thanks for your comments and suggestions, which made this book better.

Lastly, thanks to my parents and my siblings: who, despite being on the other side of the world, found ways to be near.

For the new adventure that has been bringing this book into English, I want to thank my agents at Indent for looking out for me, Alison Gore for taking a chance on this book without having read all of it, and the rest of the team at Restless Books. And to Kelsi Vanada, for being my partner throughout this whole process. I hope this is the first of many for us.

Finally, I've come to learn that books have strange journeys. I met Fabiola Menchelli because she read *The Visible Unseen* in Spanish. We worked together to make a book and became friends. Now her images adorn the second life of this book. Thank you for everything, especially for the lesson that art is an open conversation that never ends.

Thanks to the editors of *Tupelo Quarterly*, who published an earlier version of "The Act of Seeing Through" in 2019, and to the judges of the 2021 John Dryden Translation Competition, who selected "The Act of Self-Seeing" as a finalist.

ANDREA CHAPELA has a degree in chemistry from the UNAM (National Autonomous University of Mexico) and an MFA in Spanish Creative Writing from the University of Iowa. She is the recipient of multiple awards, including the José Luis Martínez National Prize for *The Visible Unseen*. Her stories have been published in the journals *Tierra Adentro*, *Este País*, and in various anthologies. In English translation, her publications include poems in *The Brooklyn Rail InTranslation* and an essay in *Tupelo Quarterly*. She was named one of *Granta*'s Best Young Spanish-Language Novelists in 2021 and lives in Mexico City.

KELSI VANADA is a poet and translator from Spanish and Swedish. Her book-length translations include *Damascus, Atlantis: Selected Poems* by Marie Silkeberg (Terra Nova Press, 2021), which was longlisted for the 2022 PEN Award for Poetry in Translation; as well as *Into Muteness* by Sergio Espinosa (Veliz Books, 2020) and *The Eligible Age* by Berta García Faet (Song Bridge Press, 2018). Her translations of Mexican writer Andrea Chapela's work have previously appeared in *Granta 155: Best of Young Spanish-Language Novelists 2*, *Tupelo Quarterly*, and *The Brooklyn Rail InTranslation*. She published *Rare Earth*, a chapbook of original poems, in 2020 (Finishing Line Press).

FABIOLA MENCHELLI has a degree in Computer-Mediated Arts from Victoria University and an MFA degree in Photography and Visual Arts from Massachusetts College of Art and Design. She has been invited to distinguished artist residencies such as Skowhegan School of Painting and Sculpture, Bemis Center for Contemporary Arts, Casa Wabi, Casa Nano, and Unlisted Projects. She has received multiple awards, including the Fulbright–García Robles Fellowship, the Acquisition Prize of the XVI Mexican Photography Biennial of the Centro de la Imagen, and is currently part of the National System of Art Creators grant from FONCA. She currently lives in Mexico City.

RESTLESS BOOKS is an independent, nonprofit publisher devoted to championing essential voices from around the world whose stories speak to us across linguistic and cultural borders. We seek extraordinary international literature for adults and young readers that feeds our restlessness: our hunger for new perspectives, passion for other cultures and languages, and eagerness to explore beyond the confines of the familiar.

Through cultural programming, we aim to celebrate immigrant writing and bring literature to underserved communities. We believe that immigrant stories are a vital component of our cultural consciousness; they help to ensure awareness of our communities, build empathy for our neighbors, and strengthen our democracy.